BRITISH COLUMBIA IN FLAMES

CLAUDIA CORNWALL

BRITISH COLUMBIA IN FLAMES

Stories from a Blazing Summer

Copyright © 2020 Claudia Cornwall

1 2 3 4 5 — 24 23 22 21 20

All rights reserved. No part of this publication may be reproduced, stored in a retrieval system or transmitted, in any form or by any means, without prior permission of the publisher or, in the case of photocopying or other reprographic copying, a licence from Access Copyright, www.accesscopyright.ca, 1-800-893-5777, info@accesscopyright.ca.

Harbour Publishing Co. Ltd.
P.O. Box 219, Madeira Park, BC, V0N 2H0
www.harbourpublishing.com

Images by Gordon Cornwall except for pages XIII (top) and XIV (bottom) by Linda Botterill; pages VI (bottom) and VII (all) by Jesaja Class; page XI (bottom) by Shawn Cropley; page XIII (bottom) and cover by Eric Depenau; page III and pages VII and IX (top) by Bailey Fuller; page I (top) by Bob Grant; page XV (bottom) by Kevin Haggkvist; page 72 by Andra Holzapfel; page IV (top) by Bryan Johns; page XIV (top) by Joanne Macaluso; page II (top) by Brad Pierro; page 219 by Brad Potter; pages X (bottom), XI (top), XII (both) by Steven Seibert; page II (bottom) by Wanda Shep; pages 117 and 119 by Samantha Smolen; page XV (top) by Wolfgang Viertel; pages IX (bottom) and X (top) by Krista Vieira; and page VIII (bottom) by Barb Woodburn.
Edited by Pam Robertson
Indexed by Ellen Hawman
Cover design by Setareh Ashrafologhalai
Text design by Carleton Wilson and Becky Pruitt MacKenney
Printed and bound in Canada

Harbour Publishing acknowledges the support of the Canada Council for the Arts, the Government of Canada, and the Province of British Columbia through the BC Arts Council.

Library and Archives Canada Cataloguing in Publication

Title: British Columbia in flames : stories from a blazing summer / Claudia Cornwall.

Names: Cornwall, Claudia Maria, author.

Description: Includes bibliographical references and index.

Identifiers: Canadiana (print) 20200183885 | Canadiana (ebook) 20200183907 | ISBN 9781550178944 (softcover) | ISBN 9781550178951 (HTML)

Subjects: LCSH: Wildfires—British Columbia. | LCSH: Wildfires—Social aspects—British Columbia.

Classification: LCC SD421.34.C3 C67 2020 | DDC 363.37/9—dc23

To the people of the Cariboo-Chilcotin, whose courage and resourcefulness are so inspiring.

What matters most is how well you walk through the fire.
—Charles Bukowski

Fences were a common casualty during the fires. Over a thousand kilometres of them burned.

CONTENTS

INTRODUCTION 1
In for one hell of a ride

SHERIDAN LAKE

CHAPTER 1 9
We can't go south;
we can't go north

ASHCROFT

CHAPTER 2 16
Get out, it's coming!

CACHE CREEK

CHAPTER 3 24
What's the name of your dentist?

CHAPTER 4 29
We got out every hose and pump we had

16 MILE HOUSE

CHAPTER 5 36
It's scary shit!!

100 MILE HOUSE

CHAPTER 6 45
An empty house in an empty town

LAC LA HACHE

CHAPTER 7 61
Like a noose

QUESNEL

CHAPTER 8   71
Five- or six-foot waves breaking over the top of our canoe

CHAPTER 9   80
Prepare for no one and nothing to come

WILLIAMS LAKE

CHAPTER 10   93
I can't answer messages fast enough

CHAPTER 11   103
Evacuate, evacuate!

CHAPTER 12   116
Go Sam, Go

HANCEVILLE–RISKE CREEK

CHAPTER 13   129
Phenomenal heart come out

CHAPTER 14   139
The fire draws everything toward it

CHAPTER 15   146
But it came, it came, it came, it came, it came

CHAPTER 16   157
You're going to die, you need
to get out of here!

CLINTON

CHAPTER 17   169
Oh fire just take me

CHAPTER 18   179
I had to go over the mountain

CHAPTER 19   184
Our neighbours were everything to us

PRESSY LAKE–70 MILE HOUSE–GREEN LAKE

CHAPTER 20   195
What the hell was wrong with us?

CHAPTER 21   208
Will you pick up Bear?

CHAPTER 22   216
I could wrap a wet towel around my head,
go out into the lake

SHERIDAN LAKE

CHAPTER 23   226
It come in here with a vengeance

CHAPTER 24   241
You get used to riding with flames

CHAPTER 25   252
Let us conspire with the forests

EPILOGUE   263
SELECT BIBLIOGRAPHY   265
ACKNOWLEDGEMENTS   268
ENDNOTES   270
INDEX   281

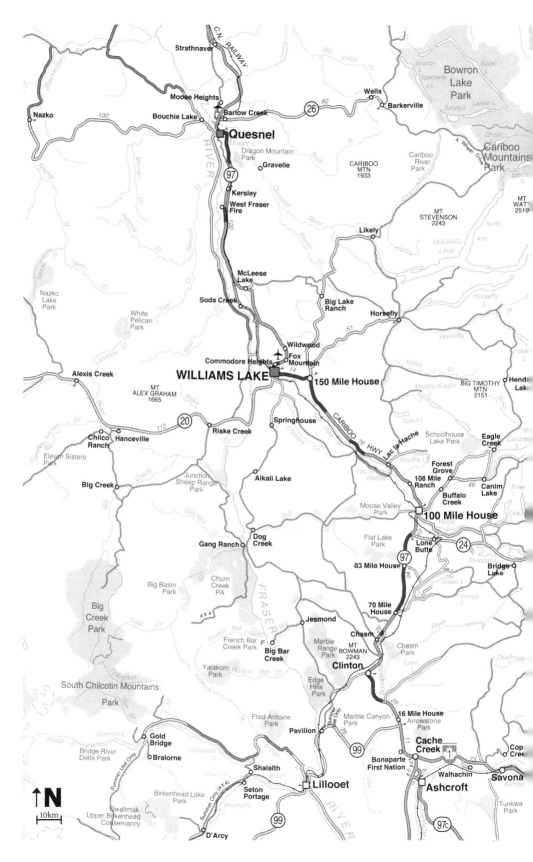

INTRODUCTION

IN FOR ONE HELL OF A RIDE

JULY 7, 2017, IS A day many British Columbians will never forget. A day like no other—unprecedented, they say. Mayor John Ranta was standing on his front lawn in Cache Creek when he saw fire aiming straight for him. Andra Holzapfel noticed smoke curling over a mountain and heading toward her just after she had unpacked her canoe in Bowron Lake Provincial Park. Flying south in a Twin Otter from Fort St. John, Jeremy Sieb noted smoke columns at Otter Lake, Dragon Mountain and south of Quesnel on the west side of the Fraser River. Chief Francis Laceese saw plumes on his way from Kamloops to the Toosey First Nation in Riske Creek. In the Walmart parking lot in Williams Lake, Raylene Poffenroth snapped photographs of several fires erupting on the horizon. Kurt Van Ember witnessed lightning strikes as he was driving west along Highway 20. "We are going to be in for one hell of a ride," he said. The fires descended like a pride of dragons, roaring and snorting. They were fast-moving and everywhere. The BC Wildfire Service was stretched to the limit and didn't even get to all the fires reported.

The Elephant Hill, Hanceville–Riske Creek and Plateau Complex fires started on July 7. They stormed across British Columbia, growing ever larger, seemingly unstoppable. Glen Burgess, an incident commander for several fires, told me during an interview at the Kamloops Fire Centre, "Ideally on a fire, you go in and you take action and you put it out. But when a fire is a hundred kilometres long by seventy kilometres wide, you can't have people on every

INTRODUCTION

Chief Francis Laceese saw plumes as he was driving from Kamloops to the Toosey Nation.

inch of that. The growth was so volatile and there were times we couldn't put our people out there. We had to let these fires do their thing. When the smoke is thick and it's heavy, you can't fly aircraft. You can't operate safely. And really it's about Mother Nature at that point." The Plateau fire was the largest single fire in the province's history. It eventually extended over 545,000 hectares and came to within 60 kilometres of Quesnel. By the time these dragons gasped their last, they had consumed a total of 1.2 million hectares—that's 1.3 percent of BC, an area more than twice the size of PEI and slightly bigger than Lebanon.[1]

SUMMER FIRES ARE A FAMILIAR occurrence in British Columbia. They have scored and shaped the province, causing devastation as well as renewal. Our natural spaces have developed through exposure to fires, and many of our plants and animals are exquisitely adapted to them. Lodgepole pines depend on heat for their seeds to germinate, a phenomenon called "pyriscence." Their cones are sealed with a resin that melts when exposed to fire and then the seeds are released. Fireweed and huckleberries flourish on the sites of burned forests, taking advantage of the extra nutrients in the ashy soil and the additional sunshine available due to tree loss.

Fire beetles fly eagerly toward forest fires to mate and deposit their eggs under the bark of smouldering trees. Dying trees are hospitable environments for the beetles because they no longer secrete the juicy resins healthy trees use to flush out invading insects.² Woodpeckers follow the beetles, attracted by their succulence. Bears, which are opportunistic omnivores, arrive somewhat later when food sources such as huckleberries get established. And deer come, lured by the lush meadows that eventually grow in the new clearings.

People, too, have used landscape fires to serve their own ends. Historians have found evidence dating back to the 1700s of First Nations in the Pacific Northwest burning for various purposes—among other things, to make clearings around their villages, to create pasture for their horses and game, and to make space for growing medicinal herbs and food. According to forester John Parminter, they relied on fires to promote the cultivation of at least eighteen different species of plants.³ Sometimes these burns escaped, but for the most part the Indigenous people had techniques to control them. The result was a patchwork of forest and grassland, resistant to catastrophic fires.

New settlers, prospectors and lumbermen also set fires, on occasion quite carelessly. In 1915, H.R. MacMillan, who was Chief Forester at the time and later one of BC's most famous lumber barons, observed:

> Forest fires began with the heroes of the Northwestern and the Hudson's Bay Company, who burned their way through Tete Jaune, down the Canoe, the Columbia, the Thompson, and through Cassiar, Cariboo, and Atlin; again in 1860 the forests blazed, lighting the way of the placer miners up the Fraser, the Kootenay, and a hundred other streams. The prospectors who uncovered the mines of the Boundary District wrecked the forests of the region. During the first sixty years every valley has felt the effects of fire; in this period the Province of British Columbia has lost by fire about seven hundred billion feet of merchantable timber, more than now exists in the whole of Canada, enough to supply the whole Canadian domestic and export demand for over one hundred years. There is no record in history of such a loss as the fire loss of British Columbia during the past two generations.⁴

To prevent such destruction, BC passed the first Bush Fire Act in 1874. It set penalties for people who let fires break away and damage Crown land or private property.⁵ When MacMillan was hired as Chief Forester in 1912, the BC Forest Branch was established to further protect the timber resources of the province.⁶ One of its mandates was to put out fires. By the early 1930s, the BC Forest Branch had put a stop to most traditional landscape burning⁷ and by the 1940s, the province was committed to fire suppression in a big way. Aided by aircraft (sometimes repurposed military planes) and a network of fire lookouts, the BC Forest Service, as it was called by then, became much better at putting out blazes. In 1995, it split into two organizations. One, still called the BC Forest Service, looked after the stewardship of our woods; the other, the BC Wildfire Service, took on the prevention and extinction of fires.⁸ With headquarters in Victoria and Kamloops, it now has six regional centres, each divided into smaller zones—thirty-two in total.

Despite all this, in 2003, BC got a wake-up call. That summer was remarkably hot and dry, and more than twenty-five hundred fires started in the Interior. Many were "interface fires"—that is, they affected communities as well as forests. Some settlements were hit for the first time. The initial interface fire broke out on July 22 at Chilco Lake and the last one on August 20 at Radium.⁹ During that fire season, the worst BC had ever seen, 334 homes were destroyed, 45,000 people were evacuated and 260,000 hectares were burned.¹⁰

In October 2003, Premier Gordon Campbell asked Gary Filmon, the former premier of Manitoba, to investigate what could be done to prevent such disasters. Filmon heard hundreds of submissions and made numerous recommendations. In his report, released in the spring of 2004, he urged BC to initiate a program to clear some of the flammable underbrush from the forest floor.¹¹ The province determined that 1.7 million hectares needed treatment. Of those, 685,000 hectares were at especially high risk of fires.¹²

On March 3, 2016, during a sitting of the BC Legislature, MLA Harry Bains questioned Steve Thompson, the Minister of Forests, about the program to remove forest debris. Thompson admitted that just 80,000 hectares had received attention. Bains persisted:

> So 80,000 out of 685,000, and we are talking about since 2004. You're looking at 12 years to treat 8 percent—rough and dirty—of what was considered to be high risk by the Filmon report. We are going year by year, and 12 years later, we're still sitting at 8 percent. If you go at this rate, you're looking at 100 years to fix this. I mean, that's the reality.[13]

In 2017, slightly over a year after this exchange, a fire season began, the likes of which British Columbia had never seen. Aside from government inaction on the Filmon report, a number of factors were to blame. Global warming had come to the Cariboo during the twentieth century and average temperatures had increased by 1°C.[14] This shift was enough to provoke a cascade of damaging effects that made fires much more likely and more dangerous. Hot, arid summers created tinder-dry forests. Higher temperatures caused more powerful storms, which blew down more trees, which in turn provided more ready fuel. They also increased the likelihood of lightning. Every 1°C of warming boosts the number of lightning strikes by 12 percent.[15] And, of course, strong winds are capable of fanning flames into raging infernos that are difficult to extinguish. As incident commander Glen Burgess put it, "We cannot stop these fires, not at that point. You have to pull back."

In addition, the changing climate created favourable conditions for the mountain pine beetle epidemic, which began in the early 2000s, peaked in 2004 and lasted until 2012.[16] Before that, spells of -40°C winter weather lasting several weeks had kept the beetle population in check. But when those frigid periods no longer occurred, the beetles spread over 18 million hectares, and left half our pines dead in their wake.[17] Though some of the dead trees were salvaged and removed, most remained in place, adding even more to the fuel load.[18]

BC's forest management practices compounded the problems created by global warming. Our zealous fire suppression efforts since the 1940s had unintended consequences, as forests became denser and flammable brush accumulated on the ground.

A number of scientists have been documenting changes in the incidence of fires in BC by looking at the stories that tree rings tell. René Alfaro writes, "Dating of fire scars in the Cariboo-Chilcotin Plateau of central British Columbia indicated that fires were much less frequent in the 20th century than they were in the 19th century."[19] Lori Daniels and Wesley Brooke studied the Alex Fraser Forest in the 140 Mile area. They examined tree rings going back to 1619, when King James I was on the English throne. Twenty-two fires occurred from 1619 to 1943, but none after that.[20] The accumulation of incendiary materials increased the probability that when fires did break out they would be larger, hotter and more aggressive. In 2017, forests monitored by both the Cariboo and the Kamloops Fire Centres had, on average, three times more fuel available for combustion than they had from 1996 to 2005, and well above the levels in 2003 and 2010, in themselves notable fire seasons.[21]

In addition, BC's silvaculture has relied heavily on one species. "Single-species planting of lodgepole pine (*Pinus contorta* var. *latifolia*) following clear-cut logging or wildfire has been common throughout interior British Columbia," forestry consultant Jean Roach writes. Pines are fast-growing and desirable so they were cultivated even outside their traditional range.[22] Because they were planted after clear-cuts, the resulting stands were mostly uniform in age. Years of fire suppression ensured that these cohorts grew to be quite mature, but the older they were, the more vulnerable they were to the pine beetle. Many trees became susceptible all at the same time. Yes, global warming contributed to the epidemic, but so did clear-cutting and species selection. We helped to create the conditions needed for an outbreak, which in turn put our woods in danger of mega-fires.

On July 7, 2017, the stage was set. It only takes one spark to start a blaze, and a plethora of sparks hit BC when a ferocious thunderstorm cut a wide swath through the province's Interior. A total of 138 fires started that day, and at 9:30 in the evening, Todd

Stone, the Minister of Transportation, declared a provincial state of emergency.

LIKE MANY BRITISH COLUMBIANS, I was glued to the news about the fires. I worried about our summer place at Sheridan Lake, which had been in my husband's family for sixty years. Some people's houses were destroyed half an hour after they first saw flames, and I didn't think we were particularly well-prepared for fires that volatile. I was concerned about my friends and neighbours too. I knew how many of them loved their homes and cottages and how, in some families, the attachment to a ranch or lake went back for generations.

Certainly there were losses, but as the summer went on, I also came across wonderful stories about how people coped in trying circumstances. I thought we could learn lessons from them about the importance of community—apparent in so many ways. British Columbians looked after strangers who had no place to go. They helped each other rescue and shelter animals. They pooled information. I wanted to share what I had learned, so I began to collect anecdotes for a book. I liked the idea of being a conduit for the people whose voices I heard—for those who lost their homes and cottages, for those who were evacuated and displaced, for those who fought the fires raging around them, and for those who watched helplessly from the sidelines.

I spoke to ranchers, cottagers, First Nations people, RCMP members, evacuees, store and resort owners, search and rescue officers, firefighters, mayors, park rangers, pilots and volunteers. I was often able to see events from several perspectives. And I discovered the as-yet-untold story of how Williams Lake was saved.

Many people allowed me to have glimpses into their lives and I saw how they met disaster head-on. They were brave, resourceful and kind. Among other things, this book is a tribute to their generous spirit.

SHERIDAN LAKE

CHAPTER 1

WE CAN'T GO SOUTH; WE CAN'T GO NORTH

FRIDAY, JULY 7, THE CARIBOO. "Have you heard about the fires?" Sandra McLeod asked, calling us on our hands-free car phone as we drove west on Highway 24.

"Heard about them? We just *saw* one at Little Fort," I said. "A thick column of smoke north of the highway. Gordon took a picture of it a few minutes ago." Sandra was worried about Jane Fisher, her aunt, who was on the trip with us. We were heading to our cabin on Sheridan Lake, about 490 kilometres north of Vancouver. Jane was an old friend of Gordon's mother; they originally bought the property together and we continued to share the use of it. Despite being ninety-one years old and having poor vision, Jane was always game to visit. But sometimes I think Sandra feels we're a bit too adventurous for her. We were nearing the end of a five-and-a-half-hour drive on our annual migration north. And like the birds that preceded us, we were not easily diverted from our purpose.

Sandra talked about a blaze at Ashcroft, how it closed Highway 97, and also about another fire near 100 Mile House. "That's forty kilometres away from the lake. We'll be fine," I said.

"Be safe," Sandra urged.

"We'll phone when we arrive. And I'll keep on top of things by watching the internet," I assured her.

A few minutes later our son Tom called. He is rarely anxious about anything. "Have you heard about the fires?" he asked.

"Sandra just called," I told him.

"One fire information officer said people were calling the fires in faster than he could write them down," Tom reported.

"Well, there are no fires here," I said, looking out at a peaceful grove of poplar and spruce and a hayfield where harvesting and baling had already begun. "We'll keep you posted."

Though I sounded fairly calm on the phone, the calls were rattling me. "When we get to Three Corners we should gas up," I said. Gordon smiled indulgently. He saw that my worrywart nature was starting to kick in.

At the Esso station, it was clear that we were not the only people to have the same idea. Cars and trucks were lined up at every pump. We pulled up to one and were third in the queue. The fellow at the head of the line was filling not only his pickup but a collection of jerry cans as well. When *he* was done, the man immediately ahead of us seemed to have trouble with his credit card; he went to the cashier to see about it, leaving his car beside the pump, making it unavailable to anyone else. I sighed with impatience. Gordon backed out and swung the car around to another pump. We were still second in line. *The driver in front of us must have a large gas tank*, I thought; she seemed to be taking forever. I began to wonder whether the unusual demand would deplete the station's fuel supplies before we could buy any. But the woman at the pump finally finished and we were able to fill up. I felt better with a full tank of gas.

We trundled along the dirt road into the resort at the southwest end of Sheridan Lake. As usual, we parked the car and picked up our boat, which Chris Brown, the resort owner, left tied up at the dock. Unusually, I looked down the lake with apprehension. It was a bit smoky, probably because of the fire at 100 Mile. We loaded the boat and, although it was sometimes temperamental, it mercifully started without any trouble. Fifteen minutes later, we were at our own dock.

As I walked up the path to the cottage, I saw blowdown, or what foresters call "windthrow." The trees were toppled in a storm that had ripped through the BC Interior at the end of May. It hit our region, the Interlakes area, especially hard. I knew from my daughter, Talia, and her partner, Pat Kane, who had visited the place in June, that our structures—the cabin, the boathouse, the woodshed and outhouse—had survived, and for that I was grateful. But the forest around us had suffered. The gusts, which had reached 90 kph, had snapped the trunks of even hundred-year-old trees, and uprooted others. The way to our neighbour's cottage was impassable. Trees were down on either side of the boathouse and immediately behind the cabin. Farther back, a tangle of massive trunks was scattered in the forest like a giant's game of pickup sticks. Some of the downed trees had been "standing dead": pines still erect, although already killed by the beetle infestation that had come through a few years earlier. But many seemed perfectly healthy. I had never seen carnage like this before.

The lake was always a place of refuge. While there, we returned to an earlier and somewhat simpler lifestyle. We cooked on a hundred-year-old New Brunswick-built Fawcett wood stove and heated with a large stone fireplace. The cabin was off-grid, so our lights were solar powered and we cooled our beer in a spring where the water measured around 5°C. In the summer, we swam toward a tiny islet, just offshore. A pair of loons nested there, and on the May long weekend, we'd watch intently to see whether they would produce a hatchling or two, becoming anxious when the local eagle showed too much interest. Last year, our pair produced two loonlings, but one disappeared. The eagle probably got it.

We've come up in the winter as well. One year, when we didn't have much snow and the surface of the ice on the lake was smooth, Gordon and I ski-sailed with Tom and Talia. Two of us held the ends of a tarp and used it as a primitive spinnaker to let the wind blow us down the lake. Another time, the snow was dense and packable so we made an igloo. Gordon found a shard of ice, which we used as a window. At the cabin, we *play* in ways that aren't possible in the city. We stretch our brains, look at sunsets, observe constellations, count shooting stars. Gordon has been coming here since 1958 and we had our honeymoon at the lake in 1971. We brought our kids along when they were just a few months old; Tom and his wife, Leanna, had *their* honeymoon here too. Memories flood my mind wherever I look.

We go to the lake to get away from problems. To have them turning up on our doorstep made me sad. We had the pine beetle epidemic, and then the blowdown. Not only were these events destructive in themselves, but they sowed the seeds for future destruction by providing a source of ready fuel. Would we see a fire now—a trifecta of harms?

July 8. For breakfast, we had apple pancakes, bacon and several cups of coffee, boiled cowboy-style in an old enamel pot on the wood stove. Afterward, Gordon went outside to make a dint in the blowdown. He wanted to saw some of the dead trees into rounds and split them for stove wood. Jane washed up and, as I had promised Sandra, I consulted Google. Though our cabin was somewhat remote, we now had the internet via our cell phones—a mixed blessing.

On the BC Wildfire Twitter feed, I learned that around noon the previous day, as we were having lunch on our drive up, evacuation orders were issued for properties at 105 Mile and 108 Mile House. While we were crossing the lake to our cabin in the late afternoon, Cache Creek was evacuated. When we were assessing the effects of the windstorm on our place, the airport at Williams Lake caught fire. I discovered that the Gustafsen Lake fire, near 100 Mile House, had ballooned to 3,200 hectares and was 0 percent contained. At Little Fort, lightning had ignited not one but three fires—two northeast of the community, near Dunn Lake, and one to the southwest, at Thuya Lake. All were still burning and the residents had been ordered to leave.

The BC Wildfire Service was straining to cope even though it is a large organization with significant resources. Every summer, it hires a thousand firefighters and has over a hundred companies around the province on standby to provide additional personnel if needed. The service has elite "smokejumpers," who can reach sites by parachute, and "rapattack" units, which rappel down from helicopters to attack fires inaccessible by road. It also has a fleet of aircraft: tankers that drop retardant, bombers that skim water from lakes and rivers, and helicopters that can lob retardant, foam or water. And yet all of this was not enough to combat the firestorm Mother Nature had unleashed.

As I turned off my phone, my stomach churned. I went to help Gordon stack the wood he was cutting, but throughout the day I kept checking in with Google. The news did not get better. At noon, 182 fires were burning across the province and smoke was visible from many areas. By evening, Drive BC told me another section of Highway 97, north of 100 Mile House, was now closed and more of the highway was shut down south of us, between Clinton and Cache Creek. These restrictions not only prevented us from driving home via the Fraser Canyon but also blocked our access to Highway 99, an alternative route south through Lillooet, Pemberton and Whistler.

I didn't like being hemmed in. We couldn't go south, we couldn't go north. At the moment, Highway 24 was open, so we could drive east, but that could change anytime. The Little Fort fires at the eastern end remained uncontrolled. As much as I enjoyed being at the lake, I didn't want to be *stuck* there. Over dinner, I broached the idea of leaving. "We just got here!" Gordon protested. He had plans for our visit. He wanted to install a couple of new solar panels that we'd just bought and, of course, cut more wood. This made perfect sense. If a fire came, it was better to have less fuel on the forest floor. But I worried about the risks of staying. And I thought about Jane. What if she had some kind of medical emergency and we couldn't get to the hospital because of a road closure? What if we were ordered to leave at night? That would be doable, but difficult. Packing up, getting Jane safely down to the dock and navigating in darkness with an unreliable boat would be challenging. Plus there was our cat, Charlie, who was along on the trip, another complicating factor. If we suddenly had to go, I felt certain he would pick that moment to disappear on some mysterious cat business. But if we knew we were leaving, we could prepare by capturing him when he showed up for a meal.

We couldn't move on a dime. It was better for us to go when the sun was up, at a time that suited us, not to have to rush off when an emergency was imminent. I asked Jane what she thought. She said she was okay either way. She was comfortable about staying, but was also willing to go. "I'll leave the decision to you two," she said. I looked at Gordon. Reluctantly, he deferred to me. Unless there was a dramatic reversal in conditions the next morning, we would return to North Vancouver.

July 9. We woke to a lurid dawn sun and smoky air. Gordon took a picture. Google told me the evacuation order for Cache Creek had expanded to include more properties, and Loon Lake, a forty-five-minute drive northeast of Cache Creek, was placed on evacuation alert. We were not changing our plans. We closed up the cottage and got into the boat. As we pulled away from shore I looked back, wondering if I would ever see the place again. There were so many fires across the province, and the crews couldn't fight all of them. They would have to prioritize—and a sparsely populated lakeshore with no services would be low on the list.

Traffic was light on Highway 24 and no one was lined up at the gas station. As we neared Little Fort, we saw that the towering column of smoke was gone, but a thick pall lay over the road starting about five kilometres west of the village. By the time we pulled into Barriere on Highway 5, the site of a famous fire in 2003, the smoke had cleared, and the rest of our trip was uneventful.

July 10. Back home, we learned that 100 Mile had been evacuated at 8:47 the previous evening.[23] The wildfire was within one kilometre of the town. Highway 24 was now closed to all westbound vehicles and Sheridan Lake was on evacuation alert. When I called the resort to find out how things were there, Chris's partner Cindy told me that the stores at 100 Mile House, normally supplied by trucks driving up the Cariboo Highway, had started to run out of fresh food.

Gordon was reconciled to what we had done: "I think it was the right decision."

ASHCROFT

CHAPTER 2

GET OUT, IT'S COMING!

FOR A LONG TIME NO one knew what ignited that small blaze on land belonging to the Ashcroft First Nation. Lightning wasn't the cause, as the sky was clear on July 6, 2017, the day it started. The days before were free of electrical disturbances too. Nor was it likely that a spark from a passing CN train was responsible as the fire originated above a railway tunnel, a good distance away from the tracks. Early in October 2017, I talked to Josh White, chief of the Ashcroft Volunteer Fire Department. "I don't know whether this was an intentional start or not," he said. "We've had multiple fires in that area in years past. Where the fire started there wasn't much sagebrush, just grass, because the sagebrush has been burnt off so many times. We don't know what this was. I don't like to think it was criminal because I don't like thinking that way. Was it intentional? I don't know. But we know it was human-caused." Did a vehicle's exhaust set something on fire? Was it a carelessly tossed cigarette? A campfire that got out of hand?

In September 2019, after a complex investigation that lasted over two years, the RCMP announced that it had solved the mystery of what started one of the worst fires in BC's history. As I write this, the RCMP is working with Crown prosecutors to see whether charges should be laid. We still don't know the names of the persons involved.

As I spoke with Josh over lunch in the bustling local Tim Hortons, dishes clattered and conversations ebbed and flowed around us. He told me that he got a call from the Ashcroft Nation around

A grateful supporter hugs Josh White, Ashcroft fire chief.

9:00 in the evening. "Basically I phoned it in to Forestry [the BC Wildfire Service] that night and gave them all the details. The fire was inaccessible to our engine," he said.

During our long chat I learned that Josh has lived in Ashcroft ever since he was eleven, and he joined the local firefighters in 1998. I also found out that the fire was the third in a string of traumatic events he'd experienced over a period of five months. In March, he was the first person to enter the home of a family in the Venables Valley, just south of Ashcroft, where the parents and two children had died due to carbon monoxide poisoning. Then in May, after searching every day for nearly a month, he finally discovered the body of Clayton Cassidy, the Cache Creek fire chief, who had been swept away during a flood. Now this. "Psychologically, it's been a hard year," he admitted.

None of Josh's engines are equipped with four-wheel drive and the July 6 fire was on rough terrain. "So I took Forestry and guided them onto the site with my personal vehicle, which has four-wheel drive," Josh said. "They arrived shortly after midnight on the seventh and worked hard. As soon as they began hitting it with water that night, they suppressed the fire. I thought, 'This is handled.' At the time, it was only about one hectare in size—very small. The smoke column was going straight vertical, which means no wind. It was a very slow-moving, non-aggressive fire."

The BC Wildfire Service fights an average of two thousand wildfires each year and aims to contain all of them by 10 a.m. the day after they are reported. It prides itself on succeeding with 94 percent of them.[24] When Josh woke up the next morning, he looked south and felt relieved to see blue sky and not a whiff of smoke anywhere. At first, it seemed as if he had been part of another one of those success stories. But by 11:00, the temperature had soared to 34°C and the winds had picked up. Stoked by gusts from the south blowing between 50 and 60 kph, the mild fire of the previous evening had metastasized. "Must have been an ember got into the grasses and it exploded from there," Josh said.

Ashcroft is drier than anywhere else in Canada except for the high Arctic. June, for instance, averages a total of just twenty-six millimetres of rain.[25] But in 2017 it was even drier than usual. By July 7, not a drop of rain had fallen for thirty-five days.[26] The parched desert hills were a powder keg.

Just after 11:30 a.m., Josh received another call from the Ashcroft Nation. A conflagration was now threatening their homes. Josh and his crew responded right away to save what they could. "When I first arrived in the rescue truck," Josh remembered, "the fire was about two hundred feet away to the south. I got one cap off the hydrant, and I turned around and the fire was at the rear end of the fire truck. We retreated; we had to retreat three times off the reserve before we finally stood our ground, because the fire was so aggressive. I thought Forestry had it in the bag, but the next day, the fire was like a volcano. We've had countless fires on the reserve. We've nailed them every time. But this time, the way the wind direction was pushing it up the hill, everything worked in its favour and not in ours."

That day, Dale Lyon, one of Josh's captains, was driving a "tender," a truck designed to transport water to a fire scene. When I talked to him and his wife Maggie on the phone, he told me that he was on-site "with a couple of younger fellows." He said, "We were going from house to house, literally removing fuel. We knew this fire was still spreading in every direction and we thought, 'Well we can do this. We can water things down, so we can keep on moving.' By the time I got to the third house, one of the younger fellows tapped me on the shoulder and said, 'Look, look behind you.' The first house we had cleared was in flames. Yeah, like totally, and we're only talking minutes."

At that point, the BC Wildfire Service sent in air support armed with fire retardant. The firefighters on the ground got out of its way. But their work was not done. Around 1:30 in the afternoon, Josh got a call from his fire dispatch in Kamloops. Bradner Farms, a dairy on the north side of town, was ablaze. Tanks of coolant were exploding and a barn was burning. Apparently about twenty people worked on the farm and some of them had been trying to get the cows out of the burning barn. "The cows are trained to go into the building to be milked two or three times a day," Josh explained. "It's all automated. The cows kept wanting to go back in the building because it was milking time." How the people were faring was unknown.

Josh sent Steve Aie, his brother-in-law, to help out at the dairy with a crew of four in the newest vehicle, Engine 3. Moments after they had taken off, Josh realized that his dispatcher was trying to call him again. But by then, Josh said, "A repeater site had burned up on the hillside, so now we didn't have radio communications with our dispatcher. I called them on the cell phone, and they said, 'Pull your truck that's just left out of there because it's dangerous. Everyone has reported out of the dairy farm now. Everyone's safe. Get your crews out of there. It's a real dangerous area to be going in.' Right away, I tried calling them [the firemen on Engine 3] on the radio. But my repeater's down, so I can't talk to my crews either. On the top of the hillside, I looked up and there's my repeater site on top of Elephant Hill. It's just a ball of fire. I'm like, 'Okay we've lost our radio communications.' When I was on the phone with dispatch, I lost cell service too. I said, 'We've got to get those guys out of there. We've got to tell them.'" For Josh, this moment was the worst during the entire fire. "I was afraid, not for my own life but for my guys. I sent them down the road into goodness knows what. Getting that call from dispatch saying, 'Don't send them there, that place is a death trap,' when I've already sent them, and then losing communications at the key moment, as I was trying to call them, was horrible."

Josh and several other firefighters took off down Highway 1 to catch up with Engine 3. On the way, Josh said, "We almost met our maker because some guy in a semi had dropped his trailers right there on the highway. And through the thick of the smoke, we couldn't see them. We came through the smoke and there's this abandoned trailer, and we swerved around it—on a blind corner. It was a crazy scary time."

When Josh and the other volunteers reached the road leading up to the farm, the smoke cleared a bit and they could see Engine 3 heading toward them. "They made a good call," Josh said to me. "We were happy to see that."

When I phoned Steve Aie to get his recollections of the event, he said, "We were fine at first and then saw fire on both sides of the road. The road was full of smoke. You could feel the heat. The whole hill was on fire. The barn was fully engulfed. It was frightening to see. We turned around. We all decided we're not going in. In fifteen years of firefighting, this was the fastest fire I've ever seen."

While all this was happening, another flank of the fire was veering west over Elephant Hill, across Highway 97c, into Bonaparte Canyon and toward Boston Flats Trailer Park, home to seventy-six people. Sergeant Cat Thain, detachment commander of the Ashcroft RCMP, told me that one of her members had a home near Boston Flats and a couple of dogs. He'd decided to check on his dogs, and when he realized how fast the fire was moving, he grabbed them and drove into Boston Flats to warn the residents. "Get out. It's coming!" he broadcast from his vehicle. While he was on his PA system, one of the other members of the Ashcroft RCMP called him to say, "Your house is gone." His place had been destroyed in just twenty minutes. "And you know, he didn't miss a day's work after that," Cat said later.

Josh, who was driving back to Ashcroft from Bradner Farms, also saw what was happening in Boston Flats. By then the fire was so close to the trailer park that all he could do was drive in and make sure everyone had left. He didn't know that an RCMP officer had already been through, so he yelled into his PA system, "Get out. We can't do anything for you. This is going to be bad. You need to leave now." Thankfully, the residents had all evacuated and were safe. Josh sped away, getting himself out ten minutes before the fire hit. It consumed all but one trailer. The trailer park's occupants were lucky to get out with their lives. Because it had only one road in and out, they could have very easily have been trapped.

Once Josh was back on the highway, he met his fire crews and told them, "This is it. We have to go back to Ashcroft. Without cell service and without our repeater, we need to be in our community." Although Josh had a satellite communication system to fall back on, it only worked when the crews were in close proximity. "For about four or five hours, through the afternoon, we went defensive: I had a crew patrolling the downtown, along the river shore; we had pedestrians watching to make sure the fire wasn't jumping; and then we had another crew patrolling north Ashcroft along the highway to make sure the flank wasn't coming within striking distance of the town."

The ten days that followed were long, Josh recalled. "Basically, what I was doing was taking the rescue truck—I'd pull into my driveway somewhere around 12:00 a.m. and 1:00 a.m., get out of the coveralls, go to bed, put them on at 6:00 a.m. and back out I'd go." Josh and the other volunteers were constantly patrolling, extinguishing small blazes to make sure the fire did not reignite. In total, during the ten-day period when the firefighting was most intense, the eighteen members of the Ashcroft Volunteer Fire Department worked thirteen hundred hours, often on top of eight-hour shifts at their regular jobs. "We're paid $7 an hour to go fight fire," Josh said. "Our firefighting budget for the town normally is around $7,000 for a year. When we got done, it was $13,000 for the month of July. We just blew that budget out of the water."[27]

Twelve of the Ashcroft Nation's thirty-two houses succumbed on July 7; altogether, twenty-seven people lost their homes. All the structures in the village of Ashcroft were spared, but the town went without power for thirty hours and had no phone service for four days. This created a number of fairly serious inconveniences.

The gas pumps didn't work, for instance. And even when electricity was restored, you needed cash in hand to fill up. Without telephone lines, cards and bank machines didn't work and tellers couldn't access their systems. "You could have a million dollars in your bank account, you couldn't get it," Josh pointed out. Ironically, Ashcroft was never evacuated. Without electricity, the evacuation orders could not be printed.

No people died but Josh recalled, "We found a couple of dogs afterwards that didn't make it. We tried to bury as many as we could find. We made graves, so people didn't have to see that when they came home. One of the saddest images I ever saw in my life was on the reserve. Some horses were running; they were on fire, their tails were burning. But they survived."

One of the strangest images I heard about was from Maggie Lyon, Fire Captain Dale Lyon's wife. Early on during the fire, she had a hose out and was watching for hot spots on their property. "From where I was, I could see the fire going on the upper part of the hills and up Elephant Hill. And then the other finger came down and started coming right along the river and burning all that fuel. So there were times I could see it and there were other times where the smoke blew across the river. We were smoked out and couldn't see too much. At one point, I could hear a train and I couldn't figure out what the train was and then the smoke cleared. It was the Rocky Mountaineer [which operates luxury rail 'cruises' in BC and Alberta]. They had reached CN but the Rocky Mountaineer was not notified. So they came through and, yes, when the smoke cleared there was the train. They were actually driving along the tracks when there was fire on both sides."

CACHE CREEK

CHAPTER 3

WHAT'S THE NAME OF YOUR DENTIST?

CACHE CREEK SITS AT THE junction of Highway 97 and the Trans-Canada Highway. With its many motels, gas stations and fast food restaurants catering to drivers, it looks like the kind of town people rush through on their way somewhere else. And indeed, it has a history of that. Cache Creek was already an important junction in the days of the Cariboo Wagon Road: a post office was built in 1868 and the settlement was connected to an artery heading east, toward Kamloops Lake. Before that, First Nations travellers likely passed through an intersection in the same spot; the track that met the Cariboo Wagon Road and went east was originally a trail. Even the town's name alludes to travel. It's derived from "cache," a reserve of food or goods, the stores of supplies fur traders left for themselves as they criss-crossed the country.[28]

Transportation may be in Cache Creek's DNA, but that doesn't mean all of its residents are transients. The chief of the Cache Creek Volunteer Fire Department, Tom Moe, has lived in Cache Creek for thirty-seven years and has been with the department for twenty-five years. John Ranta, the mayor, has been in the village since 1972. He's been in office a long time too—elected for the first time twenty-nine years ago. And the Bonaparte Nation's tenure in the area stretches back at least ten thousand years.

The Elephant Hill fire had engulfed homes belonging to the Ashcroft First Nation at noon on July 7. Twenty-five minutes later, it was burning across from the Petro-Canada gas station in Cache Creek, 12.3 kilometres away. The fire had blazed northward at a rate of 29.5 kph. In a blog post, theoretical astrophysicist and science writer Ethan Siegel says wildfires can travel through forests at up to 9.6 kph and through grasslands at up to 22 kph.[29] Even by those standards, the Elephant Hill fire was a record-breaker.

Glen Burgess did tours as an incident commander on the Elephant Hill fire and on fires near Williams Lake.

I first met Glen Burgess in February 2018. A tall man with a military bearing, he was deployed as an incident commander for eighty days in 2017. Although he couldn't comment on his strategies for specific fires, we talked generally for a couple of hours about many aspects of them, including variable winds. In BC, he said, strong winds commonly occur when a cold front passes through. The winds ahead of the front are typically out of the south-southwest, but once the front moves on, they swing around, often abruptly, and blow from the west-northwest.

Glen started as a firefighter in 1986. Over the years, he has seen many big, menacing fires. Still, from his point of view the summer of 2017 was unique. "The anomaly that I personally saw, unlike other years, was the number of times those cold fronts hit us," Glen said. "In a typical season you'd get one or two erratic cold fronts. But I think in Williams Lake, when I was there, I saw it two or three times. On the Elephant Hill, I think I saw it four different times. The fires grew because of that. At one point on the Elephant Hill, we had sustained winds for nearly forty-eight hours. On one of the lakes, fifty-knot winds were reported. Well there's nothing we can do. When there are whitecaps on some of those smaller lakes, you know we could have every airplane in the world and we're not stopping that fire."

Volunteer Fire Chief Tom Moe first heard about the fire in Cache Creek at 1:30 in the afternoon while he was at work in the Chasm Sawmill, about sixty kilometres away. He wasn't able to leave until 3:30. By then, the fire had begun moving north on both sides of Highway 97. On the west side of Cache Creek, at the small airport, three hangars were aflame. One fire truck went to help, but unfortunately the crew members had to abort the mission. The blaze was so intense they could do nothing, and two out of the three structures burned to the ground along with their contents—a plane, a car and a bus.

Tom finally got to Cache Creek at 4:15, when John Ranta was already signing an evacuation order. The worry was that the fire might encircle the town and trap the residents. Known as a colourful character, John drives a McLaren supercar in the summer months and advocates tirelessly for higher speed limits on the highways. He told me that after he issued the order, he went home: "I stood out on the front lawn. I called my wife and I said, 'Boy, look at that, sweetheart.' And coming down the hill toward Highway 1, in the middle of Cache Creek, was a fire that was burning in the sagebrush, bunch grass—visible fire—coming in our direction. I always thought my wife was a pretty calm lady. But she didn't seem that calm, that particular moment. She started loading the pictures and keepsakes into the van. And with our son Richard helping her, they got all organized to leave town.

"There were more police officers here than I've seen in my life," John continued. "They were wandering the streets, many, many of them. The RCMP knocked on my door. They said, 'There's an evacuation order for this community. You have to leave.' I said, 'I'm familiar with that. I signed it.' They said, 'Are you getting ready to leave?' I said, 'Well, I haven't really decided.' Then they asked the question: 'What's the name of your dentist?' You sort of scratch your head and you wonder why the hell do they want to know the name of my dentist? Then it comes to you. They ask in order to try and convince you of the seriousness of the situation. They're really hoping you will say, 'Why do you need to know the name of my dentist?' so that they can say, 'This is how we can identify your remains, based on your dental records.' What they are trying to do is get you to ask that question. I didn't ask that question. I knew that's what they wanted. But they assumed I wasn't leaving. They put a coloured tag on my door knocker. One colour means the people are not leaving, another colour means the people have left. They go to every house and tag each house with the appropriate ribbon tape."

Later I discovered that the police use different colours in different communities. It's a strategy to foil looters. They want to keep them guessing about which colour indicates that the residents have gone.

When I asked John why he decided to stay, he said, "It's the embers that are blown in front of a fire that cause it to grow. You can actually save a structure by making sure the embers don't accumulate adjacent to the building, either in the eavestroughs, or on the roof, or next to the chimney, or at the back of the building. If you're there with your hose and you're able to put out embers that accumulate, you can save your house." He did not feel particularly threatened, and he was confident that the McLaren sitting in his garage could escape any fire Mother Nature could throw at him. And as he put it, "During the evacuation, I felt it was my responsibility as the mayor of the community to make sure I was here to provide any leadership required in responding to the wildfire."

Tom Moe also chose to stay put. While the fire was cresting the hills around Cache Creek, an RCMP officer told him he had to get out. He said, "I'm not leaving until the fire is at my doorstep." The officer knew he couldn't do anything to dissuade him, so he just walked away.

Nine hundred people left their homes after the evacuation order was issued. Most of them went to a reception centre in Kamloops. After these departures, Tom said, Cache Creek "was basically a ghost town." Still, a few people did not go. As John Ranta explained, "When you are told by the police that you are evacuated, they can't force you out of your home. But you have to keep to your property. You can't go wandering the streets. Obviously the concern is the potential for looting." John told me that he knew some people who remained and continued to drive around the town as they normally did. "They took the approach, 'This is my community, I pay taxes in it, I can drive around here if I want to.' They were forcibly evacuated."

The Cache Creek firefighters worked all night on the seventh. "You just run on adrenalin," Tom said. "The winds were brutal. The fire got to a waste-water treatment plant where the village had a large pile of bark mulch it had taken away from a local park. We spent four to five hours putting that out. Then fire came to the south side of Cache Creek, to the Sage & Sands Trailer Park. It got to the landlord's property and burned some trees." Working together, firefighters from Cache Creek and Ashcroft, as well as from the BC Wildfire Service, formed a defensive line at the south end of town. They stood their ground, protected the trailer park and saved the community.

In the morning, Tom went to bed. But he had no chance to rest on his laurels—or even rest much, period. After snatching a couple of hours sleep, he was back at it. He divided his crews into a day shift and a night shift. Their task was to drive around town looking for hot spots. Though the fire had arrived very quickly, it lingered when the winds died down. "It actually took two or three days for the fire to pass through—before I felt relatively safe," Tom said. As an extra precaution, structural crews arrived and put a protective three-hundred-metre strip of sprinklers at the north end of the town—in case the fire returned. But after the destruction on the seventh, Tom said, "No further structures were lost."

The crews had been able to deflect the fire away from Cache Creek, but they had not managed to put it out. For fifty-eight days, the monster would grow and grow. It would consume whatever was in its path: trees, pastures, houses, barns, fences, cars, trailers, machinery, livestock, pets and game. Many people would flee its approach, leaving their homes and possessions. Sometimes the behemoth would pause briefly, but then, energized by a fresh wind, it would rampage north again, its appetite unappeased.

CHAPTER 4

WE GOT OUT EVERY HOSE AND PUMP WE HAD

"THERE'S A FIRE AT THE Ashcroft Reserve and it's going to be in Bonaparte in no time at all!" the caller said. Ryan Day, Chief of the Bonaparte First Nation, who received the message at 12:30 p.m. on July 7, was an hour away, in Kamloops. He had been married for just two weeks. He quickly rounded up some face masks and other supplies he thought might be useful and hit the road.

We met in a coffee shop in Kamloops on a slushy afternoon in February 2018. As soon as a tall, handsome man with a shock of shiny black hair walked in, I figured it was him. I was right. My daughter-in-law, Leanna Mitchell, had put us in touch. She'd come to know Ryan when they were students at Simon Fraser University. He has worked with at-risk Indigenous youth on Vancouver's Downtown Eastside and graduated with a BA in economics from SFU and an MA in Indigenous governance from the University of Victoria. In his twenties, he was a competitive runner who won silver medals in Canadian half and full marathons and qualified for the world championships. When we spoke, he told me a little wistfully that he had less time to run than he used to have. Ryan had been elected chief in 2015 and, as well as his political duties, he had a seven-year-old and a ten-week-old baby claiming his attention.

CACHE CREEK

Bonaparte chief Ryan Day relied on the experience of band "old-timers" when fire threatened their land.

When Ryan arrived back at Bonaparte, five kilometres north of Cache Creek, the band members already knew about the looming danger. I asked him if the RCMP helped with the evacuation, but he said, "No, it was way too fast. There was not enough time for them to become involved." About sixty of the band's 280 members elected to stay and fight the fire. Stories by Canadian Press, News 1130 and other news media reported that the Bonaparte Nation had disobeyed an evacuation order,[30] but according to Ryan, the news stories weren't accurate. Nations can determine for themselves whether they are going to evacuate or not. When the band council chose to defend its land with those members who wanted to do so, it was not defying anyone; it was within its legal rights.

"We've never been faced with something like this before. It wasn't a decision we took lightly," Ryan maintained. "It wasn't a foolish decision, it was calculated. Our people knew the land, the roads, the winds." Many First Nations have a deep understanding of fires due to their traditions of using them to help game and edible plants flourish. Amy Christianson, a Métis social scientist with the Canadian Fire Service, explained in a phone conversation how Indigenous people picked their moments for fires: "If the snow was receding up a hill, they would burn up to the snow line, and the fire would extinguish itself on the snow. Then they would go in two weeks again as the snow had receded further, and burn the next stretch that had opened up. They would burn at regular intervals. In some areas, they might burn every two years, in others, every ten."

Several members of the Bonaparte First Nation had twenty years of firefighting experience under their belts and a distinguished teacher, Percy Minnabarriet, had schooled a number of them. Not only was Percy inducted into the BC Cowboy Hall of Fame in 2008 for the many rodeo competitions he won, but he was also a famous firefighter. In the spring, he would visit the local high school and ask the students if they wanted to make money. "They'd skip out of school in April or May, and that would be it, they'd learn how to fight fires," Ryan said. Percy was dedicated, knew what he was doing, and was tough as nails. In *Wildfire Wars*, journalist Keith Keller described how Percy reacted when his leg was crushed by a bulldozer:

> They'd managed to keep him in hospital at Kamloops for a year or so, but after he got out he repeatedly frustrated his doctor by cutting his hip-length cast to below the knee so he could get back to riding horses. Finally the doctor sealed him in steel rods, but they only lasted until Minnabarriet had [his wife] Marie pick him up a new blade for his hacksaw.[31]

AS RYAN DESCRIBED GETTING READY for the fire, he said, "We had the knowledge sitting there waiting, not enough equipment. We got out all the hoses that we did have. We had an emergency plan but we hadn't practised or anything. We don't have a fire truck, we don't have really anything. We have a number of hoses, which we strung together. Our reservoir was small and we had a new reservoir built but it wasn't online yet. We had the old reservoir, and it would run full blast for twenty minutes and that would be it. So it was a one-shot deal. But anyway we got organized and got the hoses and pumps out."

After the preparations were completed, Ryan said, "We just waited. It probably took less than a half-hour before it got on the doorstep. We got to a critical point where the fire was coming straight into the community. A band member had some drip torches and brought those out. We lit our own back burn that stopped the fire from coming in and burning that first house. Once the first house caught, that would have been it. It would have jumped house to house, building to building. It would have gone through the whole community. It was critical to prevent it from getting to the first house. The wind was just right, we lit up the grass and it burned straight up the hill and created a guard for the community. At the same time, one of our band members with a skid steer [a piece of machinery that can pull, push or lift materials] made a firebreak to join the back burn to the highway."

The fire blew up the west side of Cache Creek, came to the road crossing Bonaparte land, but didn't burn over it. It paused, seemed to hesitate and then Ryan saw a "little bit of smoke" on the hill on the east side of Highway 97. He said, "They were doing bombers in Cache Creek at that time. It only would have taken one dump on that and we wouldn't have burned the whole forest. I remember I was standing there, thinking we should run up there and just stomp that out. Of course we didn't because we were protecting our houses. We were just looking at it. It was not an easy road to get there, but a couple of trucks with water tanks could have taken care of that."

Was Ryan right? Could a simple intervention at this point have prevented the destruction that was to come? Armchair incident commanders will no doubt have views about the situation. It is hard to know. On July 7, the BC Wildfire Service was overextended already, with 138 fires breaking out. In hindsight it may be easy to identify the ideal response, quite another to do it in the heat of action. In any case, the brief moment of possibility passed. Ryan turned his focus to the task at hand—saving the buildings under threat.

In the early evening, at 5:30 or 6:00, the Bonaparte First Nation asked the Cache Creek Fire Department if it could spare one of its trucks. Fortunately, Cache Creek had reached a brief hiatus in its own battle with the fire, which for a time was boxed into a canyon. It was able to fulfill the request. Later, around 4:00 in the morning on July 8, the BC Wildfire Service arrived to provide more help. "Some forest service guys got on-site and came and put out little hot spots on the edge of the hill," Ryan said. "They came in a water truck and sprayed to make sure there was no possibility for flare-ups next to the houses. But the fire was still raging up the hill from us. The whole hillside opposite, the whole mountainside on the east side of the highway, was totally on fire. We kept an eye on it and kept rotating shifts through the night."

The fire came within a hundred metres of the first house, and the community lost a derelict building that hadn't been used in years. But it saved forty homes, as well as the hall, the church, a preschool, the office and the brand new water treatment system. One save was quite a surprise. The fire burned all the way around one particular house and when Ryan peered at it through the dark smoke and shooting flames, he was sure it was gone. But then the air cleared and he could see the house was still standing. Ryan attributes this turn of events to the homeowner's diligence. She was scrupulous about clearing away brush and weeds.

"The thinking was we absolutely have got to defend these houses," Ryan said, "because it takes forever to rebuild houses through the Indian Act. It just gives you that much more motivation to protect them because you know if you don't, people are going to be out of a home for a year or more." In August 2017, with an idea of streamlining operations, the federal government replaced the agency serving First Nations. The Department of Indigenous and Northern Affairs Canada was dissolved and two new departments were created: Crown-Indigenous Relations and Northern Affairs Canada, and Indigenous Services Canada.[32] Whether the change was enough to speed up the grind of bureaucracy remains to be seen. Two years later, the Ashcroft First Nation is still waiting for new homes to replace the twelve lost to the flames. "It's slow going any way you cut it," Ryan maintained.

Even for the Bonaparte community, complete recovery will take time. Two-fifths of its watershed burned. Fireguards created during the fires made the land easily accessible to cars, trucks, 4 × 4s—and unwanted visitors. Ryan has been holding monthly meetings with other affected First Nations communities and provincial forest managers to develop "fire salvage principles"—ways to ensure the land has the best chance to restore itself. "The ecosystem is so fragile, you have to be careful about what you're going to do," he said. "Everybody is so worried about the economy and making sure the mills stay running, but if you do things irresponsibly, you can add another decade or two to the restoration of the forest."

You don't get over a fire as big as the Elephant Hill fire in a hurry. "You see some pretty extraordinary sights," Ryan recalled. "The size of the flames at night, the plumes of smoke that look like mushroom clouds, like nuclear bombs. And now when I drive through a valley and there is no wind and people are burning slash and the smoke is sitting in the valley, I'm back there instantly—that smell. It reminds you right away."

But being "back there" is not always a bad thing. "We earned the respect of the firefighting crews that came in and everybody involved because we stopped our community from getting wiped out. That was recognized across the province," Ryan said. I could hear the pride in his voice. "Certainly, we would have been in trouble without the skills and knowledge we had.

"We knew what our limitations were and did the best with what we had. That was enough, fortunately."

16 MILE HOUSE

CHAPTER 5

IT'S SCARY SHIT!!

ON JULY 7, BARB WOODBURN was cutting a client's hair at the Revelations Hair Salon in Ashcroft. She was watching the fire tear by along the hills above the town when she told her customer, "I need to get home." The wind howled as she raced back to her place, a small ranch off the Old Cariboo Road, west of Highway 97 at 16 Mile House, about twenty-five kilometres north of Cache Creek. It was a twenty-minute drive away. When she arrived, her husband, Rob, told her, "I'm going to work on the fire up at 100 Mile. I'm just getting ready to leave."

"You're not going anywhere," Barb replied. "Hop in the truck. Let's go down the road." Only fifteen minutes away, the Bonaparte community was surrounded by a wall of flames. Rob took one look: "Holy crap," he said. He made a phone call, cancelled work, and then he and Barb began helping their beleaguered neighbours load up their livestock. "They actually moved a lot of their animals up to our house," Barb remembered.

Two days later, an RCMP officer arrived at the Woodburn place and said, "You have twenty minutes to get out. That fire is coming over the hill." But they weren't leaving, at least not yet. They had to make sure all their animals were safe. In addition to the livestock for which they were providing a refuge, Barb and Rob had their own five horses, fifteen ewes, three rams, chickens and a couple of dogs. They also had a problematic steer that wasn't halter broken and had never travelled in a trailer. Because her daughter in Dawson Creek had their horse trailer, Barb said, "We had no transportation, nothing to get these animals out." But word got around about their predicament and friends from the Bonaparte First Nation showed up. They loaded up their own horses and the Woodburns', as well as the challenging steer, and moved the lot to the Perry Ranch, out on the Trans-Canada Highway past Cache Creek.

In the meantime, Barb put the ewes in her truck. "They were sardined in," she said. Joining a long line of vehicles under police escort, she drove south to the Perry Ranch. People waved to her as she passed, as if she were in some kind of parade. This made her laugh—but only for a minute. And then the reality of the situation hit her and she burst into tears. All she wanted to do was get back home. That was now more complicated than usual. A checkpoint had been set up near Hat Creek to keep people out of the fire zone. She could go south, but not north to her own ranch—not without a good reason. Fortunately, a solution appeared in the shape of the driver of a fire truck that followed her to the Perry Ranch. She told him she needed to get back. "I lied to him," she recalled. "Actually I said, 'I need to go home and get some more animals.'" The driver offered to take her into the fire hall to obtain clearance. Armed with the necessary pass, she went through the checkpoint without any problem.

"Thank God we did stay home, because we set up the water line and put water on the roof. We tested that a million times a day just to make sure. We were watering down our barn, our house, our shop—everything, right?" The Woodburns had been living on the ranch for nearly eighteen years, which wasn't particularly long by Cariboo standards. But when they moved in, there wasn't even a road; they had built everything from scratch. They had invested a great deal of work and love into their place. To think of losing it all was unbearable. "You're like, 'This fire is coming,'" Barb said. "All of a sudden, we're driving up to the highway four, five, six times a day. You would see a plume of smoke and you're just like, 'Where is it? Where is it? Where is it?'" On July 13, fire threatened some hydro lines on a nearby slope. Barb caught a dramatic shot of a plane dropping retardant on the lines and posted it on her Facebook page. "It was just like MASH 4077," she said.

The same fire was endangering nearby 16 Mile House. The community of fifty families on the east side of Highway 97 was too small to have a fire hall, fire engine or formal volunteer fighting force, but it had made plans nonetheless. "Neighbours helping neighbours" is how Allen Midgley, a long-time resident, described their arrangements. The folks at 16 Mile House had started working on fire-preparedness over ten years earlier. Roy Simpson, a community member who died in 2008, had the idea of making a mini-tender—a trailer capable of carrying a 250-gallon tank of water. The trailer could be attached to a truck and deployed as needed. A simple but practical innovation, it caught on. Other residents built their own trailers and then a local mining company, Constantia Resources, donated $20,000 to the community for a well, a water tank and a pump dedicated to firefighting. "We've protected against the fires along Highway 97," Allen told me. "If you can put them out close to the road, you have a chance. But if it gets up in the hills, all hell breaks loose. We've had eight or ten of these fires around here that we've been instrumental in controlling."

As fire was closing in on the community and the critically important hydro lines, a structural protection unit arrived. The crew set up a balloon bladder, hooked it to the communal water tank and strung hoses along the south side of 16 Mile House.

At 11:00 in the evening, the fire was so intense that an RCMP officer showed up at the Woodburns' ranch for a second time. He said, "You need to get out."

Rob said, "I'm not leaving."

And when the officer looked at Barb, she said, "Yeah, me neither."

"Okay, we're going to need the name of your dentist."

"You're joking, right?" asked Barb.

"No, I'm dead serious. When you die, we need to know who you are."

"Oh my gosh!"

As Barb said to me, "There's fear, and there's total fear—of everything. The RCMP member said, 'Now all of 16 Mile, every structure, is burnt to the ground.' And we're like, 'What do you mean every structure is burnt to the ground?' And he goes 'Gone, totally gone. Get out.'" To emphasize his point, he pulled out a photo of a building reduced to ash.

But the Woodburns stayed put and in the morning 16 Mile House was still standing. Brian Kuzyk, another long-time resident, credited the firefighters. "I believe they saved 16 Mile. They were trying to save the power lines. We live close to them so we got saved too," he told me in a phone interview.

"You know what burned?" Barb asked me. "He [the officer] had a picture of one man's shop that burned. But the rest survived. People stayed, they fought it." When she learned that, Barb took to Facebook:

> July 15
> 12:19 PM
>
> I'm hearing rumours so let me clarify. 16 mile has not lost any houses that I know of. One shop is gone. Horses are ok. I hope this helps a bit. I just want it be over!!!! It's scary shit!!

"I made it a joke," said Barb, "that I was the 16 Mile news reporter." But it actually wasn't a joke. She had a reporter's instinct and dedication to the truth. "When I read that somebody's house must be on fire because there's smoke, it drives me insane," Barb explained. "What I'm thinking is that I need to clarify, I need to be honest." She vowed not to churn the rumour mill but only pass on what she knew to be the case. "If you say that so-and-so's house might have gone up, you're putting the fear of God in everybody. You don't do that to people."

The situation started to improve. The winds eased off and temperatures went down a notch. The fire was still creeping northward on the east side of the highway, but it looked as though the Woodburns had escaped. On July 19, Barb and Rob drove to Cache Creek. "I have tons of eggs," Barb said, "and I'm thinking 'Well maybe somebody needs them in town. So I'm going to give them to the food bank.'

"We get to the roadblock and the RCMP says, 'If you cross this line you're not coming back.' So we say, 'Okay, we'll just back up and we'll go home.' They say, 'No you're here already, you gotta go ahead.' And this was after we had brought our animals home. And now we have other people's animals also, because they're getting burned out."

The Woodburns reported to the fire hall in Cache Creek, as the RCMP had told them to do. While waiting to register, they gave their eggs to some of the people who had business in the hall. There didn't seem to be anybody else to give them to. "We just wanted to help, and then it turned into such a kerfuffle," Barb said. When friends from the Bonaparte community came into the hall, Rob had an idea. "Can you get us home?" he asked them, and then he explained what he wanted to do. "That's when we lied," Barb said. Rob told the people in the hall that they were going to help some neighbours with their garbage. This was reason enough to warrant passes; the Woodburns got a couple and drove north. As their friends went west on the turn-off to Bonaparte, Barb and Rob continued on the Cariboo Highway toward the checkpoint. When I asked whether the police recognized them, Barb said, "One RCMP member was awesome. It was like he was on our side. When we flashed our little passes, he was like, thumbs-up and, you know, go, go, go. He recognized us for sure." I smiled when Barb told me the story—the "collusion" between the ranchers, the members of the First Nation and the one RCMP officer was a great example of resourcefulness.

Barb felt things were returning to normal. She and Rob unloaded their truck, into which they had packed their tack and other treasured possessions. She posted on Facebook:

July 21
8:04 AM

Well it's happened!! We are cleared to come and go. Our area is now safe. There are so many people to thank. I hope I don't miss anyone. This fire was a disaster, but was it? I believe it has brought people together. Love always in our hearts. This is Barb signing off from the 16 mile news. I will miss doing the news.

Little did Barb know that her biggest news story was yet to come. For a brief period, she relished her freedom. She organized an egg giveaway, went to the farmers' market in town and took care of her own animals as well as the extra eight "evacuee" horses—three from Loon Lake, two from Maiden Creek and three from Clinton. She did her usual farm chores, watched for the helicopters flying over and greeted the pilots with a friendly heart symbol she made with her hands. But on August 1, things took an ominous turn. The next morning, she posted:

August 2
10:21 AM

The fire jumped the hwy up by 20 mile, Scotty creek area. It is now burning on our side of the valley. This is our biggest fear for sure. Last night, we had ash, not little ash, big honking stuff. This is what causes fires. We pulled out all the hoses again. We watered the grounds and the house, and yes packed the truck up again.

An unruly wind had blown the fire west over the highway. And although Barb didn't mention it, the wind also began pushing the fire north, toward Clinton. On August 5, Jon Azpiri reported on Global TV that Elephant Hill, the province's largest single fire, had burgeoned to 110,000 hectares. "Ground crews have spent the week trying to contain the fire's north flank as unpredictable winds make it hard to make progress."[33] The wind was so capricious that it unexpectedly did an about-face and pushed the fire south toward Barb and Rob's ranch, destroying much in its wake.

Some people alleged that there was more to the story than a fickle wind: human error was also a factor. I'll return to this controversy later in the book, when you meet the Doughertys, another local couple. In the meantime, let's follow that wind.

August 5
9:33 PM

It started last night the wind was blowing at a pretty good knot. Rob and I go sit up by the Hwy to do our nightly check. Couldn't stop saying, 'omg, omg, what the hell.' Our life and everyone's life is going up in flames. Then we see the cat skinner and faller buncher punching in a fireguard where we quad or ride our horses. I honestly started to cry. I felt like the super heroes arrived. We're so happy we bombed down the back road and introduced ourselves. I almost kissed them (oh poor guys). I put my head down and just prayed. Please keep us safe.

As the smoke became thicker and the fire drew ever closer, Barb and her neighbours drew closer together too. They watched out for each other, phoned each other, shared news, shared equipment, shared hugs, shared beer. Just feeling that they weren't alone kept them sane. Knowing someone else was out there, someone who would come in a heartbeat, made a big difference.

On August 5, Barb got a call from one of the crew at 16 Mile House, asking "Do you need us?" "Well, we might not need you at this moment," she said, "but eventually we might." They both agreed to set their alarms and check in at 3:00 a.m. to see how things were going. The wind was blowing toward the Woodburns' place and unless there was another swing in direction, trouble was imminent.

Early in the morning, Rob took the call from a 16 Mile House volunteer who asked, "How's things at 3:00?"

"You know, actually it's not bad," Rob said. But fifteen minutes later, he came into the house and shook Barb awake, "Oh my God, I need your help. You better get up, the wind is howling."

When Barb went outside, she saw that the fire had come down the draw and flames were everywhere. "We need help," she said simply to Rob.

"No, I think we got this," he replied. But after seeing the look on her face, both incredulous and dismayed, he changed his mind: "Okay, go ahead, phone whoever you need to."

Barb grabbed the phone out of her pocket and right away reached the 16 Mile fire brigade. "Okay," she said, "now I need help."

The 16 Mile volunteers arrived around 3:30 a.m. Brian Kuzyk was the first to show up, trailering a 500-gallon tank of water. When she saw him, Barb had an immediate sense of relief. Then another set of headlights appeared in the darkness, and another, and another ... In total, five trucks arrived. As well as Brian's tank, they had one trailer loaded with six 250-gallon tanks and three trucks carrying single 250-gallon tanks. Barb began to feel exhilarated. *Bring it on! Hell, we can deal with anything now!* She was almost spoiling for the fight. The enemy—who had been lurking out of range, taunting her, terrifying her from a distance—was here, where she could engage. She was reminded of the song "The Devil Went Down to Georgia," in which the Devil and Johnny have a fiddling duel. If the Devil wins, he gets Johnny's soul; if Johnny wins, he gets a golden fiddle. Fire is pouring from the devil's fingertips as he pulls the bow across the strings, but Johnny still wins. And so did the Woodburns and the 16 Mile House gang. By 6:00 in the morning on August 6, the flames were subdued and Barb was handing her friends celebratory beers.

One of Barb's dogs, a little pit bull, had been on edge throughout the whole battle. As soon as that fire went past, she greeted each and every person. Barb thought it was if she was thanking them.

The fire came within fifteen metres of the Woodburns' house. It did not touch their main fence or their barn. Although it did take some their grazing land, much was saved. Finally, the 16 Mile House reporter posted on Facebook:

> August 6
> 10:50 AM
>
> Hi guys we are all safe.

100 MILE HOUSE

CHAPTER 6

AN EMPTY HOUSE IN AN EMPTY TOWN

ROGER HOLLANDER, THE FIRE CHIEF in 100 Mile House, got a call from his dispatcher at 11:20 in the morning on July 6 about a wildfire near a couple of mills at the north end of town. "My deputy, Brandon Bougie, and I responded right away," Roger told me as we spoke in his office at the fire hall in the middle of 100 Mile. The community of two thousand people straddles Highway 97 and got its start during the Cariboo Gold Rush when a roadhouse was built a hundred miles north of Lillooet.

Roger and Brandon discovered that the fire, about two hectares in size, was in an undeveloped area many people used as an unofficial gun range. As there had been no lightning that morning, they suspected a human cause—perhaps target shooting. It wasn't unusual for the 100 Mile Fire Department to respond to a wildfire in that location. But what happened afterward *was* unusual. Neither Roger nor the town of 100 Mile had seen anything like it before.

Roger called for assistance from the fire departments at 108 Mile and Lone Butte, as well as the BC Wildfire Service. The fire was on Crown land, slightly outside the 100 Mile House protection zone. It was about a kilometre away from the town's sawmills—uncomfortably close, because their assets, lumber and logs, were highly flammable. If they caught fire, the town was in deep peril. Roger recognized that he would need BC Wildfire's expertise

100 Mile House fire chief Roger Hollander points to a photo of the back burn that protected 100 Mile House from the Gustafsen Lake fire.

and equipment to combat the blaze. By 1:00 p.m., its air tankers were on the scene, and more firefighters and a helicopter were coming. A BC Wildfire command team took over the management of the incident, now called the Gustafsen Lake fire.

"I WAS ACTUALLY CAMPING AT the time," Staff Sergeant Svend Nielsen said after I asked him where he was when the fire broke out. He was head of the RCMP detachment in 100 Mile and we were speaking in his office. He had set aside an hour and a half to talk to me, but he was ready to answer a call at any moment. A tall burly man, he was dressed for work in his bulky bulletproof vest.

Svend told me that his family had been staying at a campsite on Sheridan Lake for a few days to celebrate his father's eightieth birthday. On the morning of July 6, the sky was blue and the lake lovely. But the temperature soared to 32°C, 9° hotter than normal for that time of year. And in the afternoon, when Svend looked to the northwest in the direction of 100 Mile House, he could see a huge plume of smoke. He sent a text message to his second in command, Sergeant Don McLean: "Nice little fire you

Staff Sergeant Svend Nielsen heads the RCMP detachment at 100 Mile House.

got there." Don texted him back, "So far so good. I'll let you know if we need you."

By the next morning, the fire had exploded to twelve hundred hectares. The fickle winds had changed several times since the fire erupted. Now they were blowing toward the north and, at 8:30 a.m., the Cariboo Regional District (CRD) had put 108 Mile and 105 Mile House on alert. The 100 Mile Fire Department was assisting 108 Mile House. Don McLean sent Svend an email, the subject line of which read, "I need you back right now." There was no message, but the intent was clear. Svend bolted down his breakfast, left his family at the campground and drove back to the detachment. When he arrived, it was around 10:30. Don told Svend that because the fire was so aggressive, the CRD would be upgrading the alert to an evacuation order.

Val Severin was in charge of a team of volunteers with the South Cariboo Search and Rescue (SAR) who were going to help the RCMP notify the residents. While the police are responsible for ensuring that evacuation notices are delivered, SAR teams have specialized training in finding people during emergencies and often assist. Val, a striking woman with red hair and flashing blue eyes, was an eighteen-year veteran volunteer with SAR. I met her in the SAR Hall in 100 Mile House, where she told me that the South Cariboo SAR has forty members, making it one of the largest such groups in the province. (The list of places it helped to evacuate in the summer of 2017 reads like a map of the Cariboo: 108 Mile House, 105 Mile House, 103 Mile House, Lac La Hache, the city of 100 Mile House, the city of Williams Lake, 150 Mile House, 70 Mile House,

Val Severin, a manager of South Cariboo Search and Rescue, has been with the service for eighteen years.

South Green Lake, McLeese Lake, Wildwood, Anaheim, Alexis Creek, Canim Lake and Hawkins Lake.) As expected, the CRD issued its evacuation order. It came at 11.48 a.m., when Svend was arriving at the 108 Mile House community hall, the epicentre of operations. The parking lot was abuzz with activity. Firefighters from all over the province were coming in to help the two dozen or so volunteers from the 108 Mile Fire Department, who were already on the scene. Svend said, "They were asking me, 'Where am I supposed to go?' I pointed them toward Marcel Reid, the local fire chief. Through the course of those first few hours it was unbelievable. You stand and watch the explosions that are happening when homes and properties are being lost. There was one point where I heard a pop, pop, *pop* and then I saw this whole black cloud. Around that time, the fire chief walked up to me and he said, 'You have fifteen minutes. We're going to lose the whole west and northwest side of the 108.'"

The RCMP and SAR volunteers scrambled. The community of seven hundred homes and twenty-five hundred people emptied out. The firefighters stayed behind, working doggedly, setting up sprinklers as the flames came ever closer. "I grew up in that area where the fire actually encroached into the 108," Svend said. In some ways that was good, he explained. He was familiar with the terrain: "I knew every trail, every nook and cranny." But it also meant that he took what happened personally. The fire almost burned the house of one of his close friends. When they were growing up, this pal was often at Svend's place for breakfast and sleepovers, and the two went to school together starting in kindergarten and all the way through high school. "The fire came to the edge of his door," Svend said, shaking his head. Two houses down from there, people lost a home. And he had another childhood friend whose family home was badly scorched by the fire. The siding completely melted. "I grew up on that street. It's hard to grasp. But they stopped it. Marcel and his crew basically saved the 108."

In the late afternoon, once 108 Mile was evacuated, the officers drove south back toward 100 Mile House. On the way, Corporal Brian Lamb radioed Svend: "I think there's a bit of problem by the 103 area." Svend recalled, "He was basically saying that the fire is right there, you could see it. I told him, 'Well get in there then.' We were literally dragging people out of their homes because they weren't aware that the fire was actually there. It had grown so fast and developed so quickly in that one spot. We started doing a total tactical evacuation of the 103 because of necessity. We didn't know where the fire was going to go. This is something we do on our own without any type of government authorization. We assume that there's a risk involved with an area. And so we make the decision." The CRD soon issued its own order—at 4:45 p.m.

That first night, Svend slept in his office on two mats from the gym; he used one coat for a pillow and another for a blanket. He lay down *under* his desk because the lights, which were left on all the time, were very bright. He wasn't very comfortable, squeezing his big frame under his desk, but he was able to sleep about three or four hours. On Saturday, he got up around 6 a.m. and started the day with a couple of briefings. One dealt with regular business—the usual calls, which hadn't stopped during the emergency. Svend called this "core" policing. The other handled evacuations, security—issues that arose from the fires.

Once the RCMP cleared an area, it had to ensure that no unauthorized people entered it and that any homeowners who stayed behind remained on their own property. Though the 100 Mile House detachment had six reported break-ins during the fires, most of them turned out to be legitimate in one way or another. For example, while on patrol, a couple of members stopped at one house because the front door and garage were both wide open. They investigated and found that a safe in the house was open too, but the money and valuables were still inside. A firearms safe was open as well, but in this case, the guns were gone. When one of the investigating officers told Svend about it, he asked, "What do you think happened?" The officer said, "I think they just left and forgot to close everything." The police secured the house, and when the homeowners got back, they found a card from the Mounties that read, "Please call us." They were amazed to hear what had happened. They were convinced they had locked up properly. Svend said, "They were frightened, they were fleeing."

The fire was always moving, dancing ahead, blown by the winds and propagated by embers. By Saturday, the Gustafsen Lake fire had ballooned to thirty-two hundred hectares and was approaching Highway 97 near the 103 Mile House ridge. People were still using the road, leaving the area, going south; some who had originally planned to stay in their homes had changed their minds. While this exodus was going on, the BC Wildfire Service was trying to contain the blaze. As Svend put it, "Brian [the corporal who had first noticed fire activity at the 103] was coming over the radio saying, 'The helicopter crews are hovering over and dropping water, trying to stop it from crossing the highway.' And then he added, 'They're dropping water right on the cars.' Actually he didn't say it that way. I can't really repeat how he said it," Svend remarked with a smile.

For Mitch Campsall, 100 Mile House's long-time mayor, ordering the townspeople to evacuate was the hardest thing he'd ever done.

Svend had to block off that part of the highway. And because his fifteen members had so many tasks to cover now, they couldn't take their usual time off. On the section of road near 103 Mile House, officers were stationed in pairs in their vehicles. This allowed one officer to sleep, yet still be available if his partner needed him. "It wasn't a very relaxing place for a snooze," Svend said. "The fire was two hundred metres away."

While the fire was growing, the responsibilities of the RCMP were also growing. But like many RCMP members and other first responders, this was not all Svend had to consider. He had a family in the area—his wife and four kids. On Saturday, his wife came back from the campground on Sheridan Lake to pick up some things from their home. She was planning to drive away with the kids, so Svend went over to the house to meet her. When she got out of the truck she said to Svend, "I'm so happy to see you." But when she tried to give him a hug, all he could think about was hastening his family's departure. Without saying a word, he jumped in the truck his wife had been driving and parked the trailer because his wife was not comfortable doing that. She looked at him with astonishment. Svend told her, "I haven't got time to chat. You've got to get out of here, so I don't need to think about you and the kids." She quickly grabbed some essentials, loaded up the truck, and the family left for Kamloops. The erratic and dangerous behaviour of the fires put everyone on edge.

BY SUNDAY, JULY 9, THE influx of evacuees had caused the population of 100 Mile House to swell to twenty-five hundred. In an interview in the council chambers, Mitch Campsall (who had been the mayor for nine years and a councillor for eleven years before that) told me he wasn't convinced his town would be a safe haven much longer. He said he had encouraged evacuees to go on to Kamloops or Prince George, where reception centres had also been set up. Interior Health had already taken the precaution of moving hospital patients and elderly people living in residential care away from danger.

On the afternoon of July 9, a meeting was convened in the local arena. "I thought it would be a meeting with twelve other people," Svend said. "I walk into the arena and I don't know why I didn't notice the bazillion cars in the parking lot. Well I'd been at it for two days solid you know, twenty-, twenty-one-hour days. I guess your brain shuts off." In fact, about six hundred people were in the audience, eager to know more about the fires and what was likely to happen. When it was Svend's turn, he spoke about how the RCMP was handling security. He started to take questions, and then all of a sudden he felt a tap, tap, tap on his shoulder. He looked behind him. It was Mitch, the mayor.

Svend didn't know it but the die was close to being cast. Mitch pushed away the microphone that Svend was holding. Then he leaned in and whispered, "We might be evacuating." He smiled at Svend and turned away. This meant that Svend needed to get back to the office and start planning for the eventuality. "But it's not like you can just drop the mike and leave when you're talking to six hundred people," he said. He answered a few more questions, and then handed his microphone over to one of the other speakers. Svend recalled that as he got up to go, "This poor lady is standing there asking me about her dogs. And of course this is what we were into at that point—questions about anything and everything. So I said, 'Well, walk with me.'" The two of them walked out together and while the lady was still talking to Svend about her dogs, Constable Peter Gall was bending his other ear about what they would have to do should the order be issued.

"You know the scene from one of those asteroid movies where the guy is sitting in a car and a big asteroid goes sweeping by and there's a bunch of smoke?" Svend asked. His expression was half-fearful, half-astonished. "That was me. I mean there were two plumes and they were huge at this point. And I thought, *The world is going to end*. I said to the lady, 'Sorry ma'am, I'm going to have to go.' She's like, 'Okay, I can see why.' Because of course we were in the meeting for two hours and in that time, the fires had started up again. We get in the car and Pete looks at me and he goes, 'Should I drive Code 3?' [This is where he would peel out of the parking lot with lights flashing and sirens screaming.] 'Well, no,' I said, 'You don't want to panic everybody, just drive normal.'"

MITCH HAD SPENT ALL SUNDAY thinking about whether he would have to issue an evacuation order. At the big meeting in the arena the mayor told everyone that an announcement could happen at any time. "I wear my heart on my sleeve," Mitch confided. "I got pretty emotional. I said, 'Make sure you get your bag packed, make sure it's in the car, not at the door because it is serious.' I made the comment because I was starting to get scared." When a reporter from the *100 Mile Free Press* phoned him, a few hours later, around 7:00 in the evening, she said, "I understand you guys are evacuating us." Nothing was official yet. "We have no intentions at this moment," he replied.

But the fire was now five thousand hectares in size. The winds were unfavourable, and the direction the fire was taking seemed particularly disturbing. "The fire kept trying to get around our barriers to the sawmills," Mitch told me. Forestry was one of the mainstays of the local economy and the yards had always been a welcome sign of business activity. But if the mills ignited, one scenario that local officials considered was that a fire cloud might erupt. Svend described it this way: "If the fire got over the hill and actually into the log yards it would create an ember cloud. The cloud, based on the size of the log piles, would be about five kilometres by five kilometres, so basically cover the whole town and rain fire on it—like a volcanic episode."

The technical term for what Svend called an "ember cloud" is "pyrocumulus" or "flammagenitus." Normally clouds are created when the sun heats the surface of the earth, which causes water to evaporate and rise. As the moist air hits the cooler upper atmosphere, it condenses and forms a cloud. This also happens with a pyrocumulus, but much more quickly. And because wildfires can reach temperatures of 800°C or more, the upward drafts are powerful, sometimes towering to heights of eight kilometres. A pyrocumulus generates strong and unpredictable winds that can pick up "firebrands"—burning pieces of wood—and propel them for one or two kilometres, spreading the holocaust even farther. To me the possibility sounded Biblical in its severity.

Another issue was escape routes. Highway 97 was blocked off north and south of 100 Mile due to other fires. Only one exit remained—Highway 24. When we left our cabin on July 9, I had not heard about ember clouds, but avoiding the possibility of being trapped was now a priority for me. Even Highway 24 was not totally secure. At the time, there were three fires near Little Fort, by the junction of Highway 24 and Highway 5. Though they were not blocking the road, it was easy to imagine that with a shift of wind, they could.

ROGER HOLLANDER WAS MINDFUL OF a talk he had heard just a month before at the BC Fire Chiefs Conference in Vernon. Chief Darby Allen, who had led the defence of Fort McMurray during the 2016 fire there, described the evacuation of his city—how people drove through flames to get out. Roger didn't want that to happen again. When the BC Wildfire officials told him they were recommending evacuation, he thought it best to move on it right away. He didn't want to wait until daylight and discover that new blazes were obstructing the last way out.

Town officials held one final briefing to discuss logistics. When they were ready to go, all they needed was the mayor's signature. "He teared up when I slid him that piece of paper," Roger said. "He really loves this town."

"BANG, I GRABBED THE PIECE of paper," Mitch told me. "Boy that was tough. It was probably the hardest thing I've ever had to do in my life. You know, we've never had this community evacuated. This was all new to me. I really believed that when we came back, this town was not going to be the same. That scared me. And again, I'm being honest, I really believed we were going to lose a big portion of the town." The order was issued at 8:47 p.m. Mitch called the *100 Mile Free Press*: "Yes, we are evacuating now."

"We couldn't get the printer running fast enough," he recalled. The police officers and SAR volunteers who were going to deliver the notices had to stand and wait for the sheets to come out of the machine. They'd grab a handful, hot off the press, and be gone. Five minutes later, some more people would show up, wait while the printer was cranking away, and grab another lot.

Throughout these operations, Svend's biggest concern was that they would miss someone, that something would happen and the person would die. "That was the number one thing, my greatest fear during the whole summer." He insisted that his men go to great lengths to be sure that all the residents knew what was happening. During the evacuation, a couple of his members came to one house that appeared to be empty. The neighbours, however, insisted that five minutes before the police arrived, they had seen someone inside. So the officers knocked several times and looked in the windows, but couldn't rouse anyone. One of them finally phoned Svend and asked, "What am I supposed to do?"

Svend said, "Kick the door."

"Kick the door?" the officer asked, in surprise.

"Kick the door," Svend repeated. "You've got to make sure they are gone."

"We ended up paying for a door. That was fine," Svend said, shrugging his shoulders. "At least we knew they were gone."

Val Severin told me that during the same operation, her teams did miss one person. They were going door to door in a trailer park. They knocked at one place and after receiving no answer taped a notice to the entrance. They didn't know that an elderly lady who was hard of hearing was inside. She didn't hear them. The next morning, she woke up and saw thick smoke. She looked outside, realized no one was around, and called 911. "It can happen," said Valerie. Unlike the police, the SAR teams can't search anyone's property. "But we do the best we can, within our legal limits."

One of the factors complicating the evacuation of 100 Mile House was that nearly 30 percent of its residents are over sixty-five. They weren't necessarily going to be able to jump in a car and drive off just because an evacuation order was issued. Some had mobility issues; others didn't own a vehicle. The RCMP officers and SAR members took it upon themselves not only to inform residents about the order but also to make sure they could follow it. "My teams would come in and report a residence that had an elderly lady, say, with no transportation," Val explained. "So I would hand that information to Svend and he would coordinate with his members to go to that location and arrange transportation. In some cases, they would drive them out, or they might call an ambulance if the person was in a wheelchair or couldn't transfer to an everyday vehicle. I had my teams keep moving from house to house because our goal was to share the information and we couldn't get hung up."

Despite such complexities, 100 Mile was evacuated within an hour. "I was amazed," Mitch told me. He decided to stay. "I wasn't going anywhere. I was going to stay with my fire department and my staff because this was our command centre." He felt that if he left, he would be abandoning his colleagues. "That's just not something I could do. I thought, 'Well, I'll be the guy locking the door on the way out as it's burning and flames are coming after us.'" At 3:00 in the morning, Mitch left the office. "You're driving and it's absolutely nothing. Like nobody is out. It's just an eerie thing. I'm going home to an empty house in an empty town. What a devastation that was."

FOR THE NEXT TWO WEEKS, Svend ate, slept and worked at the detachment office. "I'd get three hours of sleep a night and then get up at around 6:00 or 6:15, and grab some food that was out in the front office. Basically I'd be wearing my camouflage shorts and my Green Bay Packers t-shirt and I would live in that for hours and then finally someone would look at me and say, 'Are you going to get dressed today—like in your uniform?' And there were some days where I said, 'No,' because there were the phone calls, briefings, handling situations, you know people wanting to talk to me and I never got to that point."

Certainly, Svend seemed fully imbued with that famous get 'er done Cariboo spirit.

"There was the time," he said, "when we stopped the bus." Buses were coming up every day with RCMP members destined for Williams Lake. Svend needed twelve more people, but he wasn't getting the staff he requested at the time so he asked one of his members to stop the bus. "Stop the bus?" the officer asked, wondering if he heard correctly. "Yeah, stop the bus," Svend repeated.

When the next RCMP bus for Williams Lake appeared in front of the detachment building, the officer went out into the road, hailed the driver and climbed aboard. Pointing to the unsuspecting RCMP members, he said, "You, you, you, you, you, you, off the bus."

"I got slapped pretty hard for that one," Svend admitted. "But you know at the end the day we got through to them that 'Hey, you know, we continue to have a need here. It hasn't gone away.'"

I laughed at this story about Svend's heist. He is a wonderful raconteur and I was often amused as we talked. But he also had a serious side, which I saw when he was telling me about one of the women who worked at the detachment as a janitor. "She lost her house in the 105 area pretty much on the first day," Svend said. Thinking she would need time off work to recover from this devastating blow, Svend told her she didn't have to come in. "She came in the day 100 Mile was evacuated and said, 'I need to work.' I said

'Is that going to help you?' She goes, 'Yep, that's going to help me.'" When Svend told the story, he choked up. "I get emotional when I think about it."

Her fortitude impressed Svend. He spoke about her to the officers working with him. During the fires, many of them were from out of town and didn't know all the staff. He said, "You know that little lady you've seen walking around outside doing all the garbage? She lost her house three days ago. So put that in perspective when you're walking around joking about stuff. She wants to talk to you. She wants to hear your stories, she wants to laugh. But she lost her house. So don't be a dick.'

"I think it really helped to focus the group because it was amazing how many people would go and talk to her, just talk to her and share those stories," he told me. "It elevated the morale and kept things in perspective."

MEANWHILE, THE FIREFIGHT CONTINUED. ROGER, the fire chief, told me, "BC Wildfire Service decided to use a back burn to create what they called a super guard northwest of the mills. This was its last-resort kick at the can. The preparation and the planning that goes into these operations is immense." First, the residents in two houses close to the fire line were alerted to what was coming and were allowed in to collect some of their valuables. Then BC Wildfire sent in crews to clear the nearby trees and ordered planes to drop fire retardant around the properties. The 100 Mile Fire Department and crews who were assisting them set up sprinklers and brought in water with a tender. The BC Wildfire Service ignited the burn on July 13, Roger recalled, "about a stone's throw from here." It aimed to take advantage of a hydro line that snaked its way across a hill between the Gustafsen Lake fire and the mills. Removing fuel adjacent to the line would make the break larger and give the town additional protection. A tall, pale-grey column of smoke roiled toward the sky and, though the winds were sometimes fitful in the area, that day they blew toward the northwest, away from the town. "It went well," Roger said. "I would say that the tactic saved the town."

By July 22 the danger was over. The evacuation order was rescinded and everyone returned. Svend remembered the celebration in the arena: "One corporal had been here for two weeks. So I said, 'Just go down there and bask in the glow of what you did.' And so he goes down, walks in the door, and this woman sees him walking in the door. She bursts out crying and hugs him for five minutes. Doesn't happen every day."

"For the most part we had so much love and thanks from the public," Roger said. "It's enough to last us a lifetime."

LAC LA HACHE

CHAPTER 7

LIKE A NOOSE

ON JULY 7, AROUND 6:00 in the evening, a woman was sobbing outside Marshall's 150 Mile Store at the intersection of Highway 97 and the Likely Road. Heather Gorrell, a supervisor with BC Parks who was on her way home to Lac La Hache, stopped in at the store. She wanted to get something to drink because she hadn't had a sip of anything since leaving her house at 7:00 in the morning. The sight of the woman in tears immediately affected her. When she discovered that the woman was panicking about her three horses and her dog caught in the fire zone, she wanted to help.

The woman, whose name was Lana, told Heather that the RCMP had a checkpoint at the top of the hill, on the road to Likely. The officers would not let Lana through, so she couldn't get to her place, which was about ten kilometres past the roadblock. The fire hazard was extreme and she had no way of rescuing her animals. "Come with me," Heather said to her. "I'll get you to your animals." She was in her BC Parks truck and figured if she put on her flashing red lights, she'd get through the checkpoint for sure. Heather was the perfect person to talk about animals in trouble. She herself had two horses, nine dogs, two cats and a ferret.

Heather's day had already been very long. In the morning, when she arrived at her office in Williams Lake, she'd learned that the Cariboo Fire Centre had decided to restrict campfires. That meant she had to go to about half a dozen parks and recreation centres and post signs about the ban. In the early afternoon, about thirty kilometres from Ghost Lake, a small remote site in the Cariboo

Heather Gorrell, a BC Parks supervisor who rescued three horses caught behind the lines, is pictured here with six of her nine dogs.

Mountains Provincial Park, she ran into a huge thunderstorm.

She watched as lightning splintered a tree into a dozen pieces and as the roads started to catch fire. "I had never seen anything like it. I got chased out of there." Heather told me. She phoned the Cariboo Fire Centre to report what she'd been seeing. The centre, at Williams Lake Airport, was itself in trouble. Heather remembered that call: "They are basically saying, 'We can't talk to you. We can't do anything. We're being evacuated. We're running for our lives. The fire is on Fox Mountain. The flames are right at the edge of the runway where all our helicopters and water bombers are. Our building is right there.' They were evacuating the building but they had just enough time to tell me, before they hung up on me, that Spokin Lake, Wildwood and Deep Creek were all being evacuated."

Heather had worked for BC Parks since 2000. For seven years before that, she was with the recreation program in what was then the BC Forest Service. In that job, she had worked on a few fires herself and was a fire boss. "I was usually pretty good with fires, but when it's not a controlled situation, when you phone your lifeline, and they say, 'We're abandoning ship, we're out of here,' and they hang up on you, you say, 'Oh crap.'"

There was more bad news. The rangers in Bowron Lake Provincial Park, in Heather's jurisdiction, called to say that a fire had erupted at the end of one of the lakes in the park. They had asked paddlers who were staying in the park to move away from the fire zone. Throughout the rest of the afternoon and evening, Heather would get regular reports about how that operation was going and would relay information to her boss. Eventually he would close the park completely. Heather was juggling many balls.

As she drove back to Williams Lake, she began stopping the incoming traffic to warn about the fires ahead. From one of the drivers she heard that a woman and her baby were staying at the Ghost Lake campground. Since there weren't many ways of getting away from it and cell phone coverage in the area was either non-existent or spotty, she became quite concerned. Heather used her satellite phone to call her rangers at Bowron Lake Provincial Park. They in turn used a cell phone to alert the RCMP, who made contact with the mother. Heather also stopped in at the recreation sites on her way out, intending to inform any campers whose communities had come under evacuation order about what was happening. "If you live in these places, you need to get home," she told them. She wanted everyone to realize how ferocious the storm was, as flames were popping up everywhere. Heather spoke to around fifteen groups; some were already packing up before she suggested they go. "They were so thankful and co-operative. They hugged me. I led the pack out and then I was coming out of Likely and saw big plumes of smoke. That got my blood pumping." At Fox Mountain, close to Williams Lake, the hills were glowing red, but they were still allowing people to go through the checkpoints.

Heather was anxious to get back to her own pets, but she heard on the radio that people were needed to move livestock to the stampede grounds at Williams Lake, where they could be sheltered. She decided to pick up a stock trailer from the BC Parks works yard, check on her own animals, and then go back to Williams Lake, where she would help transport livestock to safety.

HEATHER HAD BEEN IN THE fire zone for about twelve hours. She was well aware of what it was like. Nevertheless, when Lana described her problems, Heather didn't hesitate for one minute about going back in. Her long day was about to get much, much longer. When Heather and Lana reached the checkpoint north of 150 Mile House, the RCMP officers manning it didn't want them to go past. But Heather insisted: "I'm just getting this lady's animals and I'll be right out." The RCMP reluctantly waved her through. Heather and Lana saw flames shooting up on either side of the road and the Pioneer Logging sort yard was ablaze. "But there were fire trucks, and there was still authority, so you felt pretty safe," Heather remembered. Things were burning and there were embers, but Heather could still see where she was going.

Lana told Heather she was worried about loading a horse onto a trailer, so when they reached her place, Heather agreed to give it a whirl. She took a deep breath and began leading the first horse into the trailer she had attached to her truck back at the BC Parks works yard. Leading a horse into a trailer was not something she had personally done before, and the circumstances were not ideal. She and the horse were not familiar with each other or the trailer. Her anxiety was probably evident, which wouldn't help. But without much ado, the horse was safely stowed. Then she ran into an unexpected difficulty. The ball on the truck was actually the wrong size for her hitch, and once the horse was in, weighing down the trailer, the hitch promptly disconnected. Despite the mishap, the horse wasn't hurt, and she managed to unload it. But now the trailer's tongue was on the ground—a heavy dead weight. A neighbour of Lana's, who was also evacuating, came over to help; it took about forty minutes to get the tongue off the ground and onto a jack. Then Heather had to search for the right ball. When she finally located it, she discovered that the ball already on the truck was locked on. The only way she could switch balls was to smash the lock.

After these delays, Heather finally hitched up and loaded the three horses, one quite large, as well as Lana's dog. Though their machinery had been so miserably recalcitrant, the animals were accommodating. While Heather had been working on the trailer, Lana's dog kept trying to lick her hand. "And the horses stepped in like it was magic. They have a sixth sense," Heather said. Lana was in tears as she grabbed clothes and belongings, but she succeeded in packing the essentials: hay, dog food, tack.

As she drove off and she and Lana gave each other a high five, Heather thought, *Hey, I can do this!* It was somewhere between 8:30 and 9:30 p.m. "It is pitch black. Everything is blowing around us," Heather recalled.

They got to the roadblock that she had been allowed to pass through earlier. "They were gone," she told me. "They abandoned the post. The barricades were still there, but no people." I could hear the incredulity in her voice. And then she created the scene for me: "At nighttime things are magnified. It is beautiful in a way. I've worked at fires where at night, it's almost like dancing—a ballet. But this time I wasn't in a controlled environment where all the safety measures were in place. I felt really vulnerable and scared, plus I felt responsible for this other person. And I'm in my park ranger uniform so I look like someone with authority—that people would automatically look to. So I had to step up my game and be the person."

A second stock trailer pulled in behind Heather at the checkpoint and then another four or five vehicles showed up. "Where do we go?" one of the drivers asked Heather. She said, "I don't know." All around them, trees were burning and she couldn't see an exit route. One of the fellows in the vehicles behind Heather had an acquaintance farther along the road, at the bottom of the hill, so he called him on his cell phone. He was told, "Do not come down this way. There's a house on fire and it's jumping the road. You can't get out this way." Going down the hill was the shortest way to the highway, but Heather had to find another route. Then suddenly, from the west, a car emerged out of the clouds and flames and smoke. *A white angel car*, Heather thought.

She ran over and asked, "Where did you just come from?"

"I came by Deep Creek."

"Can you get through?"

"Barely. If you're going to go, I'd go now. I don't know if you can make it through."

Heather didn't want to go back because that was not a good place to be either. She told the other drivers, "I'm going and you can follow me."

She went slowly because she was a little uneasy about the trailer hitch and hauling horses. With her red lights flashing, she pulled into another checkpoint at Mountain House. It turned out she knew the conservation officer on duty. Pointing south, he said, "Heather, you can't go that way. That would lead you to Wildwood, the subdivision that's on fire."

"What about going through here?" Heather asked, pointing to a gravel road going north.

"I don't know," the officer said. "You're on your own. I don't know what to say. It's like a noose. Everything is on fire around us. There isn't any exit point."

Heather decided that the gravel road leading north through Deep Creek was her best option. She soon saw that both sides of the road were ablaze, however, and wasn't sure she'd done the right thing. At the same time, she was trying to calm Lana down. So she talked, told stories. And then suddenly she went quiet. Lana looked at Heather and asked, "Are you all right?" Heather said, "No, I'm not. I don't know what to do. I'm really scared."

The flames were like blowtorches arcing over the truck and the trailer, shooting back and forth. Lana and Heather could feel the heat through the windows. And the horses weren't moving, which was unusual. Usually when you're trailering horses, you feel their weight shifting. Heather wondered if they were so still because they were really frightened.

"I can't turn around because I can't see through the smoke and the embers," Heather told Lana. "It's dark and the road is narrow." But continuing on the same track held risks too. "The tires might blow," she said. "They'll probably melt. There will be some loud bangs. But we'll have the rims and we'll keep going on the rims. We'll be okay. We're just going to go through. Are you okay with that?"

"I'm okay," Lana reassured her.

They drove through a wall of fire for five or ten minutes. They could see flares being sucked backward away from the side of the road. The windshield wipers melted and shrivelled up but the tires didn't blow, because the air closer to the ground was cooler.

Heather and Lana made it to Highway 97, as did the other drivers following them.

They rolled into the stampede grounds in Williams Lake around midnight. Trains of trailers were coming and going. Heather spotted a small van with two windows open; a goat was sticking its head out of one, a sheep was poking its head out of the other. "It was both comical and heartwarming," she told me. Hearing her describe the moment reminded me of the Bible story of Noah, only here a multitude of Noahs were steering fleets of mini-arks, helping animals escape not a flood, but an inferno.

Lana and Heather were warmly received. People came up to them, grabbed the hay, gave the horses water and made sure everyone was okay. Heather gave Lana a big hug and said, "I need to get home to my place, to my animals."

On her way to Lac La Hache, a journey of about sixty-five kilometres that normally took forty-five minutes, Heather passed through two roadblocks. The people manning them did not want to let her pass. She begged, imploring them, "I just want to get home to my animals." They let her through, but said, "We can't help you. From this point on, you're on your own, we're not going in to get you." Heather said, "Okay."

"The desire to get home to get to your animals, to your children if you're a parent, really makes you think at a different level. It creates that internal ... you have to get to your dependents, whether they have four legs or two. I was just driven."

It was pitch black, very ashy. Although the fire had already gone through, the trees were blowing embers. The wind was still gusting and the visibility was bad. Heather could barely see in front of herself and she knew she had quite a drive ahead of her.

One of the worst moments was going past the Indigenous community at Sugar Cane, about fifteen kilometres southeast of Williams Lake. She saw a house in the distance on fire. She felt she could still burn up and she was having difficulty breathing because the smoke was so thick. *Nobody's going to come and get me because they told me they weren't.* She felt abysmally lonely and feared that she might not make it through this time.

She phoned her sister in White Rock on the south coast. It was probably about 1:30 in the morning at that point. After several rings, her sister picked up and Heather said, "Sharon, I just need someone to talk to. Please talk. I don't care what you talk about. I just need to hear your voice." Sharon didn't realize what was going on and Heather didn't really tell her. She didn't elaborate. She said, "I'm driving and it's a bit of a difficult area." She didn't say much about the fire. She said, "I've had a long day. I just need to get home, but I need to talk. I'm a little scared. It's been rough." Sharon helped Heather calm down and focus.

About half an hour from home, Heather's mood picked up; she felt good enough to let her sister get off the phone and go back to sleep. But her tribulations were not quite over. When she was a mere five kilometres from her house, she saw cars parked on the side of the road. Another roadblock, and no one was being allowed to pass. An RCMP officer came to her window. Heather said, "I'd like to go through." She thought, *I'm in uniform; this will be a piece of cake.* But it was not a piece of cake and she became upset. At one point, she thought she might even get arrested because the discussion was getting so heated.

"Heather, you need to calm down," the officer said.

"I'm going through," Heather insisted.

"You need to calm down."

"I don't need to calm down!"

"Yes, you do."

"This is ridiculous."

"Are you with the fire?"

"Yes, I'm with the fire."

"What is your specific duty?"

"I'm getting animals."

"Your animals or other people's animals?"

"I just got other people's animals. It was so chaotic."

But the officer was not persuaded. Heather pulled over and left the truck and trailer on the shoulder. She decided to bushwhack home through the woods. The forest was generally thick but sometimes it opened up a little when she hit a cattle trail and was able to follow that for a bit. She had her cell phone with her and considered turning on the built-in flashlight, but decided against it. *The cops will see me. They'll come and get me!* she thought. That was probably unlikely, but she was tired, and not thinking entirely logically. Although it was dark, she was confident she was going the right way. She had a lot of experience in the woods and wasn't scared anymore. The sense of being on a mission made her feel free and energetic. She stumbled on reaching a gravel pit, fell and skidded down on her butt. But she was familiar with that pit and knew it meant she was close to home—indeed, just a half a kilometre away.

She got to the main forest road, walked along it, came to her frontage road and at long last arrived at her house. It was smoky but there was no fire, no ashes. The sky was glowing, but at Heather's place nothing was burning. On the hike through the woods, branches had scratched her face; nevertheless, she felt comfortable. It was wonderful to see the little light on in her house. "And the horses came nickering to me. I gave them a big schoochie on the noses." She fed them and came inside.

"The dogs were in the snoopie den." Heather remembered. She had a drink of water, didn't bother taking her clothes off and collapsed on her bed, exhausted. "I've never had nine dogs on my bed before. But every single one had to be on my bed. I didn't care. *Everyone, come on!* It was all worth it. I just held onto my dogs. I have everything from a huge hundred-pound American bulldog down to a tiny Chihuahua. I remember just hanging onto their legs, grasping them. I had never been happier to be home in my entire life. I just felt so safe."

Heather had planned to stay on her property. But she began to run out of hay for her horses. The power was out, her well wasn't working and she had to depend on her neighbours for water. On July 14, she decided to evacuate and drove with all her animals to a ranch in the Hart-Chief area, about fifteen kilometres north of Prince George. When she arrived, Heather told me, "I slept for three days." She returned home when the evacuation order for Williams Lake was rescinded and power was restored.

QUESNEL

CHAPTER 8

FIVE- OR SIX-FOOT WAVES BREAKING OVER THE TOP OF OUR CANOE

FOR ANDRA, PADDLING IN THE Bowron Lake Provincial Park was the fulfillment of a lifelong dream. Ever since she was a little girl, growing up in nearby Quesnel, she'd wanted to canoe the circuit, a 116-kilometre loop of lakes, rivers and waterways connected by portages and ringed by imposing peaks. But the experience wasn't at all what she had imagined. Her plans were disrupted—along with those of many other British Columbians.

On July 3, Andra Holzapfel and her husband, Rick, drove to Bowron Lake Provincial Park with their two sons, Tom and Jack, who were five and three. On their first night, they stayed in the Bear River Mercantile, a combination guest house, restaurant and museum. The next day, they started paddling. Andra and her husband had already completed a couple of short canoe trips on other lakes with their children; this one, which normally takes six to ten days, would be by far their most ambitious.

As they set out, the lakes were like glass under azure skies. The kids were having a good time and the family made excellent progress. Andra and Rick successfully negotiated the first three portages, the longest of which was 2.4 kilometres. They even had an oh-so-Canadian encounter with a moose. On their third day canoeing, they stopped and set up camp around 4:30 in the

The Holzapfels' canoeing trip in Bowron Lake Provincial Park turned out to be much more of an adventure than they expected.

afternoon. It was still hot, so it seemed like a good opportunity to try some manoeuvres with the boys. They wanted to show them what would happen if the canoe swamped. Rick and Andra got their sons into life jackets, put them in the canoe and tied it to a tree on shore. Then they waded out and began rocking the boat. When it was half full of water, a moose suddenly appeared and disrupted the lesson. She charged the family, retreated and charged again. On her second foray, she strode into the lake to a depth of about a foot. Andra and Rick got behind the canoe and dragged it to where the water was up past their waists. As they did that, a branch on the tree to which it was roped swayed, spooking the moose. She moved back, allowing Rick and Andra to retrieve their paddles and life jackets, cut the line and take the canoe further out into the lake. However, she returned for a third time, and the family stayed in the boat watching her for about an hour as she presided angrily over the campsite. When another party of paddlers joined them in yelling at the moose, she finally left. "It was a real adventure for the boys," Andra said.

On July 7, the Holzapfels were a little late starting because the moose had kept them up the night before. They got going around 10:00 a.m., but made good time and reached the end of Isaac Lake, the biggest in the chain, a whopping thirty-eight kilometres long. Their destination was just a little farther, a small lake, McLeary, on the south side of the circuit. To get there they went on one more paddle, a minor one, and completed a couple more portages. Although the portages were short, they were strenuous because Andra's youngest son, Jack, had fallen asleep, so she had to carry him as well as her thirteen-kilogram pack.

When they got to McLeary Lake and found a campsite, it was early, one or two in the afternoon. "But we were done for the day," Andra said. "We were tired." They wanted to rest up before attempting the next section, the Cariboo River, one of the trickiest parts of the whole trip. Then, just after the Holzapfels unpacked their canoe, they spotted a lightning strike. Almost right away, a huge plume of smoke roiled up behind them from somewhere near Isaac Lake. They saw the column getting bigger and bigger, and the smoke started curving over the mountainside toward them.

At the same time, Mark Taylor and Terri Carlson, two park operators, were in a speedboat travelling along Isaac Lake telling visitors about the newly imposed campfire ban. They and two park rangers, Jeremy Pauls and Ashley Brassington, were the staff responsible for the 150,000-hectare park. On July 7, the rangers were in the southwestern part of the circuit and the park operators in the northeastern section. Although campers were not allowed to use any motorboats, the park staff had them so they could patrol the area and offer assistance to any campers who needed it. When the lightning storm started, Mark and Terri sought shelter. "We could literally watch the storm come through and the fire start. It was pretty incredible," Mark told me. From his vantage point, he saw more clearly than the Holzapfels. "There were seven smokestacks, not just one. It was quite a sight really. You knew pretty quick that it was something pretty serious." Like Heather Gorrell, Mark called the Cariboo Fire Centre to report the blaze and also discovered that the centre itself was being evacuated. "That was a little spooky to hear. Obviously if they have been evacuated, you know that removes a lot of our support. If something did happen, something medical, it's a long way."

Without any guidance from his superiors, Mark started a local evacuation. "You didn't have to be a genius to realize that it was a good idea to get everybody out," he said. "The fires were growing fast—they got to over a thousand hectares within a couple of hours. The smoke was thick and you heard that noise, that freight-train noise that everyone talks about. And then, of course, the official word came through Heather."

The fire was burning on both sides of Isaac Lake near the southeastern end. This was more or less the halfway point in the circuit. Turning people around to go back from there meant they faced about eight kilometres of portage. "Not everyone's favourite cup of tea—portaging their canoes when they've just done that over the last two days," Mark observed. But going forward when you had a thousand hectares of forest fire ahead of you, pitch black smoke columns, and dense smoke settling in the valleys wasn't an appealing option either.

OVER BY MCLEARY LAKE, ANDRA and Rick could see there was cause for concern. They abandoned their idea of stopping, reloaded their canoe and launched on the Cariboo River despite the hazards the river posed. The park map warned: "Navigating the silt-laden waters of the Cariboo River requires care and attention. Canoeists must remain alert for sweepers, deadheads and other hazards at all times." "Sweepers" are fallen trees that block the passage of a canoe and "deadheads" are stumps or logs that are mostly or fully submerged.

"The weather started to get bad, windy," Andra recalled. Though paddlers can encounter squalls and choppy water in the middle of Isaac Lake, it was uncommon to run into that where they were. "As the fire picked up you could feel it sucking the air toward itself," Andra said. The blaze was generating its own bluster.

The Cariboo River was swift, so Andra and Rick had to keep a sharp eye out for the hazards the map had flagged. They hadn't done much river canoeing, but steered through without trouble. Five kilometres later, at the end of the river, they spotted an emergency radio. They got out of the canoe and reported the fire to the park rangers, who had heard about it shortly before their call. "The wind was kicking up and we couldn't stay on the river there any longer. So we started to go and were near the entrance of the next lake when the wind pushed up really, really strong. Five- or six-foot waves started breaking over the top of our canoe," Andra recounted.

As the water swept over them, Andra looked at her two boys and thought, *What have I done? What kind of mother am I that I'd put my kids in this situation?* But she didn't have time to pursue this line of thought much longer. The circumstances demanded her full attention. They were in danger of swamping and the steep, sharp waves made steering difficult. Aiming to go west, they found it difficult to make headway. Normally, if canoeists encounter waves they will try to point their bows across them. In general, less water will splash in the boat if the waves are coming over the bow than if they are hitting the side. The craft is also less likely to tip than if it's sideways to the waves, rocking in the troughs.

A nearby sandbank looked inviting. Andra and Rick did what you're not supposed to do. They turned sideways to the waves and let a surge of water thrust them onto the beach. They landed successfully. Within minutes, Andra had the kids out of the canoe and wrapped in a blanket. When it started to rain, she got out their rain gear. They erected an emergency shelter and Andra felt better. The kids were warm. She *had* taken care of them. They had been in proper life jackets in the canoe and she had been prepared with the appropriate equipment. She felt like she wasn't such a bad mother, after all.

Then the two park rangers, Jeremy and Ashley, approached from the western end of Lanezi Lake. The rangers were telling campers to move away from the forest fire. Andra and Rick, of course, had been trying to do that when they ended up on the sandbank. The rangers understood why they might have sought temporary shelter there, but reminded them that they were only supposed to camp in the designated sites. Makeshift spots like the sandbar were not permitted. They weren't safe and it was difficult for the rangers to keep track of people who weren't in the official camping areas. Since Andra and Rick were past the halfway point, the rangers didn't ask them to backtrack, which was a bit of luck. Though there were portages ahead, they only amounted to two kilometres altogether, so the second half of the circuit was not as arduous a prospect as the first. Fortunately, the wind had begun to drop some, allowing Andra and her family to pack up and leave. It was 5:00 or 6:00 p.m. They had been ready to quit at 2:00. "Now," said Andra, "We were *really* tired."

Anxious about tipping, Andra and Rick hugged the shore. The first campsite on Lanezi Lake was full, so they pressed on. A big group of people was already staying at the next site. As Andra related this part of the story, she said something that astonished me: "We tried to stop there but they wouldn't let us come in." I couldn't believe that when you had bone-weary parents and exhausted kids, coping with a forest fire, high winds and rough water, anyone would be so unaccommodating. "This was a guide with a whole bunch of tourists," Andra said. "Because we were slower, we ran into them several times. They were all young, in their twenties. They were taking sightseeing tours and stopping and doing little hikes. They were not taking the situation seriously at all. The rangers were really upset with them. I think that they might not be allowing the guide to take tours in there anymore." I was so surprised by the anecdote, because in the many conversations I'd had about the wildfires, this kind of heedlessness was rarely mentioned.

Rather than confronting the unfriendly group of travellers, Andra and Rick moved on. When they finally landed at campsite number 34, it was dark and they'd paddled over eighteen kilometres, completed two portages and unloaded the canoe twice. They had been on the go for about twelve hours. Andra remembered how warmly they were welcomed: "It was a real community feel with all of the people staying there. When we arrived, there were eighteen canoes, but at least two or three more groups came in after us. It was amazing how many people helped us just to get dinner ready and to set up. As more people came in, we were all sharing food and helping each other and encouraging each other and it was a real neat atmosphere at that campground."

FIVE- OR SIX-FOOT WAVES BREAKING OVER THE TOP OF OUR CANOE

OVER ON THE NORTHEASTERN PART of the chain, park operators Mark and Terri were still trying to make sure everyone was safe. The canoeing conditions were far from ideal. The fire-generated winds were stirring up whitecaps. Ash rained down, which made it hard to tell water from smoke. "When it got to be dusk, and people were still in the fire zone, we loaded them on the boat," Mark said. He gave a lift to about ten people, some of whom had sustained minor injuries. Normally, people paddle for two or three hours a day, so it wasn't surprising that making them go for eight would cause a few aches and pains. "At times like that," Mark said, "it's just easier to put them on the boat and take a bit of stress off them and it speeds things up too." Of course, there was a limit to how much the staff could hurry the evacuations, since they could offer no motorized way of negotiating the portages. Everybody had to walk those. It was about midnight when Mark and Terri finally felt their visitors were out of immediate danger and got off the water.

ANDRA REMEMBERS HOW HEAVY THE smoke was the first night of the forest fire. It was not all that far away from them and gave the gibbous moon a lurid orange cast. The next day dawned grey and smoky. It was as if an impenetrable fog had descended. "It started to feel ominous," Andra said. Oppressive, too, was the lack of sound. There were no birds or insects—at least no bugs flying through the air, only corpses of them littering the surface of the lake. Instead of all the usual summer buzzing, cheeping, peeping, squeaking, chirping and chirruping, an awful dead silence reigned. That morning, before Andra and her husband launched, the rangers told them the whole park was being evacuated. They had to get out as quickly as possible.

The family had three portages left before the end of the circuit. This was where Andra felt most uneasy because of the possibility that a fire across the portage route might cut them off from an exit. She didn't know how many fires the lightning strikes had started. Nor did she know exactly where the fires were. The rangers tried to be helpful, but they didn't have a lot of information either. One of them told Andra they were waiting for someone to do a flyover, which would give them a better picture of what was happening.

"But we were a very low priority in the province. We didn't know that at the time," Andra said. "We just knew that there was no help coming, that nobody was flying over, that nobody was looking at it. It felt like nobody really cared what was going on. We didn't know that there was this big provincial emergency going on. We felt abandoned."

"The whole time we were telling the kids, 'If you can do this, we'll go to the ghost town when we get out,' because the circuit starts near Barkerville," Andra said. Barkerville was at the centre of the Cariboo Gold Rush and is preserved as a historic site and tourist attraction. "When the kids started to get tired we just would promise them we were going to go to the ghost town and we were going to have a lot of fun. We would play games with them as we went, sing songs, try to keep their spirits up and not show them how we were feeling."

Andra and Rick were driving pretty hard to reach the end of the chain of lakes, but at one point on that fifth day, the smoke cleared a bit. They could hear birds and bugs again, a good sign. Andra and Rick were beat. They decided they had gone far enough—that they could call it quits for the day. But after they unloaded the canoe and began setting up camp, the smoke came back, and the birds and insects turned eerily quiet. In light of all the uncertainty about the location of the fires, they opted to move on after all. They took down the tents, loaded up and did the remaining two portages, pulling in at campsite 44. From there on, they would be travelling on water. Even if there were fires in the neighbouring woods, they felt they'd be safer. But the long days and physical strain were exacting a toll. "By the last little lake that we did, I was falling asleep in the canoe paddling because we were so tired," Andra told me.

On July 9, the weather was bad again. On Bowron Lake, the last lake in the circuit, the Holzapfels were thrust to shore by strong winds. "It took us a long time to find somewhere to put in. It's really cliffy there. We were stressed out about it. We found the tiniest little place where we couldn't even set up a tent and we were talking about setting up a tarp emergency shelter for the night if it didn't slow. We were only about two kilometres from the end of the circuit but we were stuck there."

They waited for the wind to die down, and then pushed to the end as fast as they could. "The whole time my husband was yelling at me, 'I don't see any whitecaps yet,' and we just kept going," Andra said. "We figured if we saw whitecaps, we'd pull onto the side. I was watching for rocks and things in the water because when there are waves it's harder to see those things. And then we were so relieved when we got in."

The Holzapfels spent one night at Bowron Lake Provincial Park campground, where they phoned their families, who had no idea how they had been faring. That was how they learned that their story was just one of many—that a maelstrom had hit BC and lightning had ignited hundreds of fires. The park rangers and operators evacuated just under 250 people. They cleared the eastern section of the park in two days, and the western portion in two and a half. Despite everyone's haste and agitation, no one had an accident. No one capsized.

A transplanted New Zealander, Mark had never been through anything like the forest fire in Bowron Lake Provincial Park, but he had served for eight years in the New Zealand navy. "Ultimately, that's good preparation for stressful situations," he remarked, chuckling.

Andra thought her kids were "real troopers" throughout the whole ordeal. "It took us five days and ten hours and we figured that is the world record for a three- and a five-year-old." Indeed, the boys' achievement put them in exclusive company. The vast majority of British Columbians who were evacuated in the summer of 2017 escaped fires by using a vehicle. They hopped into a car or truck and drove somewhere—to a friend, a relative, an evacuation centre. Only a small percentage got to safety by walking and paddling out. Of those, Tom and Jack were surely among the youngest. Remarkably, the boys wanted to come back another year, and they did. The Holzapfels paddled the circuit again in the summer of 2018, this time taking a more leisurely eight days to complete it—under perfect conditions.

CHAPTER 9

PREPARE FOR NO ONE AND NOTHING TO COME

"IT WAS LIKE AN INFERNO right off the bat. The lightning was hitting, all dry, no rain. We watched some of those big fires start from ground zero and the winds were so crazy that they just took off." Max Forester, a junior member of the BC Wildfire Parattack Unit, was describing what he saw on July 7, during a flight from Fort St. John to Williams Lake.

Midmorning of that day, the unit had received an urgent request to send crews to the Williams Lake Airport. The speciality of this elite group of firefighters, also known as "smokejumpers," is the pre-emptive strike. They mobilize swiftly, fly to a new small fire, parachute in, extinguish the blaze within forty-eight hours and charge on to the next. The parattack unit has two planes, a DC-3 based in Mackenzie and a Twin Otter in Fort St. John, always at the ready, loaded with pumps, chainsaws, fuel, over 150 metres of hose per person, stretchers and first aid kits, as well as enough food and water to sustain crews for two days. Smokejumpers can get anywhere in BC within two hours. Physically fit and rigorously trained, they operate independently and adapt to changing situations. When you're a smokejumper you never quite know what you will encounter.

Four crews with three members each headed to Williams Lake. Two of them flew down in the Twin Otter and the other two drove. As the plane headed south, the crews aboard watched a firestorm unfurl below. The Twin Otter crossed the Chetwynd Zone and passed into the McGregor Mountains east of Prince George. "I saw about fourteen or fifteen fires around Otter Lake," Jeremy Sieb said. He was a training supervisor and a "spotter" who had twenty years of experience with the smokejumpers. His tasks on a mission like this one were to assess fires, communicate with the fire centres, determine what gear the crews needed, and make sure the crews jumped safely. Jeremy radioed the Prince George Fire Centre to report the fires. "Would you like us to action any of them?" he asked, but was told not to stop and fight anything: "Carry on to Williams Lake."

In the next valley, another ten small fires. Jeremy also called them in and confirmed that his smokejumpers were not to respond to anything, but to continue on. Then the Twin Otter crossed the boundary between the Prince George and Cariboo Fire District; Jeremy switched over to the Williams Lake radio repeater links.

The first major fire they encountered in the Cariboo was about twenty-five kilometres southeast of Quesnel and just west of Dragon Mountain. It was around ten hectares in size and Jeremy described it as a rank 3 or 4. The BC Wildfire Service considers rank 3 a "moderately vigorous surface fire." Ground crews attacking one of these directly "may require air support from tankers, skimmers or helicopters dropping water or fire retardant." A rank 4 fire is "a highly vigorous surface fire." Starting to move into the crowns of trees, it may be growing by tossing embers ahead of itself. Jeremy contacted the Cariboo Fire Centre in Williams Lake to announce his discovery but the message he received was the same: "Keep going." Normally, any of the fires he had encountered would have been reason enough to stop. But this was an exceptional day.

About thirty kilometres south of Quesnel, Max said, "We saw three lightning strikes—all in a row on top of a big bank that went down into the Fraser River, right on the edge of a farmer's field." Three fires, which came to be known as the West Fraser Road complex, started almost immediately. Jeremy called to report what he was seeing beside the Fraser River, but again he was instructed to keep going.

Then, as the Twin Otter neared Williams Lake, a startling announcement blared over the radio: "We are shutting down operations. We are being evacuated ourselves." Max compared the fire centre to the "brains of our operation." But now the information, overview and guidance it usually provided were not available. Jeremy said, "Everyone in the field needed to be extra cautious because of the extreme fire behaviour and because radio communications were shutting down. It was essentially up to the crew to fly around and figure out where they might usefully deploy themselves."

The Twin Otter turned around and went back north. Jeremy wanted another look at the West Fraser Road complex of fires. Due to its proximity to private properties, it seemed like a worthy target. When the plane got to the complex, he thought the fires had already increased in intensity and he suspected that a jump would be a no-go because of high winds. But he also figured that if he didn't properly assess the possibility of jumping safely, he wasn't doing his job.

First, Jeremy and the pilot chose a "drop zone"—a spot where jumpers could land. They picked the large farmer's field near the fire. It was flat and free of hazards such as boulders or heavy slash, which can create problems in some of the more remote locations to which smokejumpers are often assigned. The next step was to gauge the "wind drift" and see whether the jump was actually feasible.

Jeremy held some red and yellow crepe-paper streamers and crouched beside the open door of the Twin Otter. When he let go of the streamers, the pilot was flying around 350 metres above the ground and had positioned the plane so it was directly over the field. The streamers mimicked the path of a smokejumper under a parachute. By watching where they landed, Jeremy could see the "wind line"—the direction in which the wind was pushing. Then the pilot made a second pass. He lined the plane up above the streamers, pointed into the wind, and flew over the drop zone. Jeremy noted the time it took for the plane to go from the streamers to the drop zone—thirty seconds. The pilot maintained his heading, and at thirty seconds past the drop zone, Jeremy let go of a second "check set" of streamers. They landed within the zone (plus or minus forty-five metres). Jeremy had confirmed that the point at which he let go of the check set of streamers was the exit point. This is where jumpers would have to leave the plane in order to hit the drop zone.

Now he could calculate the wind drift—how the force of the wind would affect the jumpers. Jeremy told me that the Twin Otter was flying about fifty yards per second.[34] Since the correct exit point was thirty seconds past the drop zone, the wind drift equalled fifteen hundred yards. This was three times the allowable limit. Jeremy explained that every hundred yards of wind drift indicates three miles an hour of wind. So that day, the wind was going at about forty-five miles an hour.[35] Most of the crew were experienced jumpers, but even for them, the wind was too much. "We would have got creamed on the ground," Max said. Jeremy aborted the mission.

Smokejumping is inherently risky. You are exiting a moving plane, aiming to get near (but not too near) an untamed treacherous fire while the wind is howling. But the risk can be managed, Jeremy told me. Since its start twenty years ago, the unit has conducted nine thousand jumps, with only one injury and no fatalities—an enviable safety record.

Jeremy directed the pilot to go farther north so he could look at the Dragon Mountain fire again. Max said, "We found a spot in the trees to jump there and we threw streamers on that one and it was about seven hundred yards of drift, which we can jump. But it's kind of the upper limit of what we can do. The spot was pretty small." The opening was about a quarter of an acre in size and at those wind speeds, not safe. Since the plane was now close to Quesnel Regional Airport, where the zone fire office was located, Jeremy decided to fly in and come up with a new plan.

THE TWIN OTTER LANDED BETWEEN 5:00 and 6:00 p.m. "It was like a war scene," Max said. "You could just see these big columns of fire outside of town to the southwest and southeast. And nobody was at the base. The Blackwater Unit crew that is normally there was gone. I believe there were two wildfire assistants. They were overwhelmed and nobody really knew what was happening or how many fires there were. We were told to just start moving all these hay bales off the airstrip. The Cariboo Fire Centre had been evacuated and dozens of helicopters were going to be leaving. So we made them some spots to land and as soon as we finished rolling all these hay bales out of the way, it was like *Apocalypse Now*. A fleet of helicopters was flying through the air. I don't know how many, quite a few."

A crew of smokejumpers—Tom Hutchinson, Wren Dirks and their leader, Ryland Bennett—grabbed the gear from the Twin Otter and boarded one of the newly arrived choppers. Jeremy and Craig Wilson, who had seven years with the BC smokejumpers and four years with a rappel crew in Australia, got on another one. Max and Nathan Stewart, who were on Craig's crew, were left behind at the base. Max said, "We wanted to get out and help because we could see all these fires going off everywhere. We were worried we were going to get stuck there for the whole day." They sat and waited, hoping that someone would remember them. And then, because it was around dinner time, they ordered a pizza.

When I asked Max why he and Nathan were seemingly overlooked, he wasn't exactly sure, but he told me that the other jumpers were highly experienced and could run complicated fires by themselves. They got sent out to figure out what was happening. He was a "newbie." It was only his second year in wildfire. But he said that didn't explain why "Stewy" was left behind, as he was in his ninth year.

The two helicopters carried the smokejumpers to the west side of the river. By then heavy equipment operators, who had been dispatched prior to the shutdown of the Cariboo Fire Centre, had arrived and were starting to build a guard. Ryland and his crew stayed to work alongside them while Jeremy and Craig decided to see what was happening on the other side of the river. The Fraser was about a half-kilometre wide at that point, and no one anticipated that the fire would be able to cross it very quickly. But as Craig and Jeremy were circling overhead in the chopper, they saw the fire storming up a hill toward an enclave of eight or nine houses on the east side of the river. They also noticed that a group of air tankers had come in and were preparing to drop fire retardant. Craig asked, "Do we have anybody over on this side?" Jeremy said, "No, they're all on the other side." Craig offered, "I can get out, manage the aircraft, make sure that everything's going in the right places and try and bring some control to the chaos." Jeremy agreed with this plan, said that he was going to return to Quesnel, and promised to get Craig more resources.

Craig hopped out of the helicopter alone and found himself in the midst of a busy scene. Two RCMP cars were parked on the highway and the officers were keeping the residents out of their houses. While the tankers were dropping retardant, the locals were on the road, milling about anxiously, fearful about their properties and desperate to retrieve their possessions and their animals. "I think a lot of people out there can't get insured properly if they've built their home and have a wood fireplace," Craig said. "I remember a guy coming up to me and being like, 'The wheels alone on that truck are worth like $3,000.' And I'm like, 'Well it's not safe for you to go in there.'" Craig wasn't sure about the rules for evacuating homeowners and didn't try to order them out. But he told them what he honestly believed and the residents respected his opinion. They developed a friendly relationship. When Craig spotted a quad that might make his life easier, the locals gave him the phone number of the owner who immediately gave Craig permission to borrow his machine. "No problem! Use it," he said.

MEANWHILE, AT THE QUESNEL AIRPORT one of the wildfire assistants came rushing to the fire zone office. "He was frantic and looked wildly around the room," Max said. "He saw Stewy and myself sitting there and he seemed happy to see us. 'You two, get as much gear as you can,' he said." The assistant gave Max and Nathan the call sign of the helicopter they were to board. "You're gonna go save some houses," he instructed them. Max asked, "Is there any more information?" The wildfire assistant shrugged. Max and Nathan went to see if the Blackwater Unit crew's warehouse had anything they could use. "It had already been pilfered. There were no sprinklers for home protection. There were some hand tools that were broken and in bad shape. And there was a little bit of jewelry—things that we attach to hoses, water feeds, three-way valves. There was no hose left. No pumps. So basically we were dead in the water. But we got as much as we could—just anything we could use."

A pilot took Max and Stewy to another farmer's field on the east side of the Fraser River. They weren't told who would be on the ground with them and had no idea what was going on. "We were worried it was just going to be the two of us trying to put the fire out," Max said. "We were starting to wonder what was happening." In part, this lack of direction may have been due to the general craziness of the day. But it wasn't that unusual in the smokejumping business. "In my crew, we have a saying that you have to be fluid, flexible is too rigid," Craig said. "I always goof around with the guys on my crew that we do fluidity training. I don't tell them exactly what's going on and they just have to get by. You have to get used to it. Fires are so dynamic and everything changes so fast. If you get too attached to one idea or one plan, then you won't have the ability to shift and change with mission changes."

Although Max and Nathan didn't realize it, Craig was expecting them, and when he saw their helicopter land, he rode over on his quad to meet them. "We were so happy to see him," Max said. They gave Craig the remains of their pizza, which he was delighted to receive as he hadn't eaten since breakfast. "Wilson left me at the highway with a big group of people that were pretty worried that they were going to lose their homes. I stayed with them and was trying to talk to them and reassure them. I told them that we were going to try to do everything we could, which at that point really wasn't that much. The fire was burning up the hill pretty rapidly. It actually made it over the hill and was burning on the flat. The fire was really moving and had quite a bit of energy behind it. There wasn't a ton we could do."

Craig jumped on the quad with Nathan and they did some reconnoitering together. Craig recalled, "We had a look at all the residences and tried to find out where the flanks of the fire were. It was a fairly confusing setup because we didn't know where the houses were. There are places that are off the grid and scattered in the woods. I was trying to build a map and understand where everything was." Craig wanted to see how many houses were still standing, what the fuel was like and whether the retardant was working and was going to keep the fire from advancing. This was an area of hobby farms, some of them stocked with somewhat unusual animals, like llamas and miniature horses. "The llamas had eaten a lot of the grass, a lot of the fine fuel that would spread the fire," Craig said. "So it's like, 'Oh, we actually might not be in that bad a shape here.'" Not only were the llamas useful, but they were also a welcome source of comic relief. "I saw two llamas; I bumped into them in the woods, and they're both looking at me funny and I'm looking at them funny and then I hike past and then I see this tanker come by and drop retardant. I'm like, 'Oh no!' And I see these two red llamas come running out." Craig laughed at the memory.

At 8:00 or 9:00 in the evening, two trucks arrived carrying a couple of parattack crews and more equipment. One crew was assigned to the west side of the river, the other stayed on the east. Craig had a plan for what everyone should do. Communicating it went smoothly, thanks to what turned out to be a serendipitous acquisition—the pizza box. Craig made it into his operations board by drawing a map of the fire and writing down his tactics on it, with arrows indicating the proposed line of attack.

Then an unexpected visitor appeared: an official from the Cariboo Fire Centre driving an Initial Attack Unit truck. He was working in the field, had been checking out another incident and came over to see the complex of fires near the river, not realizing that any firefighters were already at the site. The smokejumpers were huddled over their pizza box, trying to figure out what to do. Startled to see them, he asked, "What are you guys doing here?"

Craig was equally surprised, "What are *you* doing here?"

"You guys must be the parattack guys."

"Yeah."

"What's going on?"

Craig told him that he had a plan to save all the houses. He just needed a few more resources. "Well I'm going to be honest with you," the official said. "None of that's coming because we have no way of talking to anybody to get it. Your priority is to keep the highway open because at this point the southern evacuation route from Williams Lake has been shut and they are evacuating north." Sure enough, ten minutes later, a convoy of cars heading to Prince George started driving through, making the official's point. He didn't stay, but left Craig in charge of protecting the highway.

Craig was disappointed to hear that he couldn't count on any more help, but then there was another unexpected arrival. Max said, "I just remember this long line of headlights bobbing in the pitch black." One fellow ran out of a truck and said, "Hey, we're from Alkali Lake and we're here to help you."

That they came at all happened more or less by accident, as Gord Chipman told me. He runs Alkali Resource Management, a logging company that looks after the timber assets of the Esk'etemc First Nation (Alkali Lake Indian Band). It's one of many companies that has a contract with the BC Wildfire Service to supply firefighters should the need arise. "On July 7," Gord explained, "everything was raging by dinner time. We knew we were going to be sent to a fire. But we didn't know which one. I called my crews and I said, 'Why don't you come to town and then we'll figure out where you're going to go.'" They met at the Williams Lake stampede grounds, a central location for everybody. Sitting in a circle of lawn chairs, wreathed in smoke that had drifted in from the many fires around, the Alkali Lake crews ate the pizza Gord had ordered.

Near 8:00 in the evening, a fire official appeared with rather cryptic instructions. He said, "Okay you guys, head north until you see fire. Stop there, and start putting it out." A big fire was ripping through a suburb called Wildwood, just north of Williams Lake, near the airport. As instructed, the crews from Alkali Lake did, indeed, drive north, but they didn't stop at Wildwood. "It was just mass confusion," Gord said. They kept going north for another hour, until they spotted a fire roaring around some houses. The crew decided this was where they were going to fight.

"Awesome!" Craig exclaimed when he realized he had another twenty firefighters. His chances of saving the houses had just improved dramatically. He held a briefing—with the help of the pizza box, of course. Craig said, "They loved it. It definitely helped a whole lot to show them where everything was, what our objective was and how the piece of the puzzle they were going to do fit into the whole picture. They knew they were helping the community and their pride and their work ethic just blew me away. We had just such a good team dynamic."

Craig wanted his crews to execute a classic firefighting manoeuvre called "anchoring in." The fire, which was moving up the hill from west to east, was in the crowns of the trees and too intense for the firefighters to hit at the head, on the eastern side. If they'd tried that, it would have bubbled out the side and run up the hill again. Instead, Craig divided his people into two groups, one for the north side of the fire and the other for the south side. Working with hoses, the two teams started at the bottom of the hill and gradually extinguished the fire from the outside in. They met in the middle, moved up the hill, and repeated this until eventually they had a wet line all the way up the hill. Once they got to the top of the hill, the fuel became sparser. "It was skulking around like a smouldering ground fire," Craig said. When he was reasonably sure that nothing below was going to attack, he asked the crews to pinch off the head of the fire. For good measure, Craig ran a safety belt of charged hose around the entire fire zone. He attached the west side of the hose to a pump fed by the river and the east side to a water tender with a big bladder. "If the winds picked up the next day, we were 100 percent ready."

The firefighters worked through the night and then Craig let the Alkali crew catch some sleep. He didn't want everyone to be asleep, so he kept the smokejumpers on duty. However, as they had been up for twenty-four hours straight, he did let them take a short break. They had no food and very little water, just Red Bulls and coffee they made by shaking coffee powder into their water bottles. Exhausted, they sat down on a road, stretched their legs and sipped their coffees.

Craig called the Cariboo Fire Centre. By the morning of July 8, it was running again, allowing him to provide an update on his activities. Craig asked, "I'm wondering when we're getting people to come in to relieve us." He was told, "Basically prepare for no one and nothing to come. We have over a hundred homes threatened, so don't expect to get anything more. Do what you need to do." Some people might have been distressed to be left on their own like that. But not Craig. "That was a relief to me. I knew that I could just get the job done properly." He understood from the conversation that if he needed to push through without sleep, he could. He kept the smokejumpers going through the day until he felt confident he had the fire contained. He said, "I guess after a while, you learn when a fire will move and when it won't. There by the river you couldn't really get better conditions for it to settle down at night."

Finally, after the smokejumpers had worked for thirty-six hours straight, Jeremy drove down from Quesnel in a truck, picked them up, and took them to a hotel so they could get something to eat and go to sleep. "I remember we had a really hard time checking in because we were all so exhausted our brains weren't working," Max said. "We were confusing the poor people at the front desk."

Craig stayed with Max and Nathan and the Alkali group at the West Fraser Road fire for about a week. The rest of the smokejumpers went on to other fires. "It was probably the most fulfilling and most challenging fire of my career. It just seemed like everybody showed up at a time that worked perfectly. There wasn't any waiting around. Everything just seemed to work really fluidly. We had an amazing time with those guys from Alkali. It was probably day five or six of the fire when the Alkali crew had someone from their community die while they were working on the fire. We were doing our morning briefing and they asked if they could do a ceremony right on the fire line. They said a few words; they were proud to be there doing what they were doing and sorry that they weren't at home. Part of it was in English—I think to help me understand what was going on. The actual singing was in their language. It was amazing, really touching.

"Everybody that lived there was so good to deal with. Once you told them it wasn't safe for them there wasn't an argument or a discussion. They trusted us and I was really grateful for that. We had real purpose. I felt invested in that fire and cared a lot. Sometimes we fight a fire that's four hundred kilometres from anything and you know that it might not make the biggest difference. We do work hard at it but you don't always know. This fire was so black and white. This was important. We were doing the right thing and we were servicing the community. It was intense, but in a good way."

A combination of hard work and good luck carried the day. At its peak, the fire grew to over 120 hectares. None of the houses in the little community were lost, although some outbuildings and sheds burned. Max told me that the llamas who were dyed red by retardant survived too, and after a week, they got their white coats back.

WILLIAMS LAKE

CHAPTER 10

I CAN'T ANSWER MESSAGES FAST ENOUGH

IT STARTED ON JULY 6, with an idea and a Facebook post. When Lana Shields learned about the Gustafsen Lake fire near 100 Mile House, she immediately began thinking about the horses. "Honestly," she said to me, "I love animals a lot more than I love people. And I really wanted to make sure that the horses were going to be okay." Lana—a pretty woman with chestnut hair, a fringe of bangs and blue eyes—lives in Williams Lake. At the time she was the caretaker at the stampede grounds. She texted Tim Rolf and Court Smith, two directors of the Williams Lake Stampede Association, to ask if they would support her idea: "I said, 'I would really like to open up the stampede grounds for people to bring their horses so that they're safe. Would you be okay with that?' And they were fantastic: 'Oh yeah, absolutely. Just do whatever you need to do.'" She posted on Facebook:

> July 6
> 7:20 PM
>
> Anyone needing a place for their horses due to the 100 Mile fire, please call me. We have pens at the Williams Lake Stampede Grounds for you.

The next morning, Lana went to work as usual, in the office of Parallel Wood Products. "All of a sudden," she said, "we had this storm and there were lightning strikes everywhere and I thought, 'Oh this is not good.' I'm very involved with the horse community and I knew there was going to be a problem quite quickly." With the blessing of her employer, she left early to get some pens made up on the stampede grounds. The facility, home to the annual Williams Lake Stampede and the Indoor Rodeo, has stables, a race track, and indoor and outdoor arenas. It was an ideal place to house refugee horses. "Of course," Lana said, "everything just went so fast, and all of a sudden we started filling up and I was like, 'You know, this is gonna be a lot more than what we can probably handle.'"

In the afternoon, at 3:13 p.m., Lana posted a message that she marked, "Urgent Please Share." People were starting to come to the stampede grounds with their horses. She realized that some folks were not only looking for a place of refuge, they also needed help getting their animals to Williams Lake. She put out an appeal for assistance with transport and asked for donations of hay and the loan of water tubs and buckets. The message netted 24 comments and 273 shares. She was well on her way. From then on, the messages flew in and out, fast and furious.

At 4:28 p.m., Lana appealed for someone with a Class 1 licence who could drive a stock trailer for people on Fox Mountain and the community of Wildwood. Ten minutes later she posted that she had lots of horses coming in and needed organizers. Half an hour after that, she reported that Chris on Fox Mountain was asking for assistance with two horses and "other critters." At 5:56, she was forwarding an offer from Kristen who was on the grounds and could haul two horses. At 6:24, Lana said she had more trailers on standby. "Call my cell," she posted. "I can't answer messages fast enough." At 7:29, she asked for more panels to make up pens. (Panels are portable fences that handlers can use to set up portable corrals for their animals.) At 7:54, she wrote that Janice needed a ride for one horse and asked, "Can someone please call her. I don't know if the road is closed," and then a few minutes later she wrote that Janice had a ride—"Thank-you, whoever you are!" At 8:36, she was looking for more volunteers to direct traffic, unload horses, feed, water and clean up pens. "Lots of tired folks down here!" And then at 8:59, "We desperately need more panels at the grounds. We have too many horses and not enough paddocks." Denise told her she had some panels loaded in her trailer but couldn't get through. At 9:01, Lana called for more buckets and Elizabeth replied, "I'm down at the grounds, where do you want them?" At 9:09, Lana asked for help watering and feeding horses. At 10:25 p.m., she was telling people to park on the track, and "please bring your own panels if you have them." At midnight, she was searching for Tricia who was missing and then Penny explained that Tricia was in the 140 Mile House area where the power was out.

Lana told me that she hardly slept for four days. She had help from about a dozen volunteers during the daytime but not at night. She sat on the couch in her trailer where she could see the people bringing in their horses; often they would wake her. "I had a lady banging on my door at 3:00 a.m. and I opened the door and she's sobbing because she had a three-horse trailer and five horses and she had to decide which three she could bring out." It was a common problem. "We had one lady who had nine horses and a three-horse trailer," said Lana. "I myself have four horses and a three-horse trailer." (When it came time to move her horses, Lana pulled the dividers out, crammed all the horses in and shut the door.)

> July 8
> 6:23 AM
>
> I am just jumping in the shower. (I haven't stayed up all night and slept in my clothes since I was 18.) Anyone willing to come down and help out feeding and watering horses it would be appreciated so much. Even I haven't had this many horses to feed all at once before.
>
> Thank you 🖤

At 7:20 a.m., Lana informed everyone that the washrooms on the grounds were open and that she had put some toiletries out for people to use. At 8:17, she was thanking Blaine for a donation of coffee and offering it to anyone who needed it. At 8:22, she was letting folks know that she had had over a hundred horses to look after. So far she was good for water tubs and panels but, she warned, "That could change in an instant." Then the GMC car dealership, Safeway and someone called Cathy donated food, which Lana offered to volunteers and firefighters. At 10:26, Lana was thanking Royce, Tim and Beaver Valley Feeds for bringing in hay, buckets and other supplies. Hay was, of course, a constant need. A hundred horses can eat roughly a thousand kilograms (two large rounds) a day.

By the afternoon of July 8, the number of horses had tripled. Lana was still able to take in more, but she was starting to become uneasy about them. She had been a firefighter more than a decade before, and she didn't like what she was seeing. "I had a gut feeling that things would get worse before they got better." She didn't think that Williams Lake would be safe for long. She thought it, too, would probably be evacuated and, if that happened, how was she going to get the horses out all at once? Besides, during an evacuation, the traffic was much slower than usual and hard on the horses. Going somewhere with a horse wasn't like putting your family dog in the back of a car. You could give a dog a bite to eat, a drink of water. You could let it out for short break, but that wasn't possible with a horse in a trailer. She phoned her friend Chelsea Wallach, in Prince George: "I said to her, 'Hey you know what? We've got over three hundred horses here now.'" Chelsea told her there was space in Prince George and offered to ask her friends in the rodeo world—barrel racers and ropers—if they would be willing to form a convoy and trailer some horses. At 6:16 p.m., Lana posted, "Just got this message from some wonderful friends in PG for anyone who is headed that way with horses. 'We are all set at PG Agriplex. The arena and rodeo grounds.'"

At 7:50 p.m., Lana was reassuring people that she had lots of room on the grounds. "Please come if you need help and a safe place for your horses." At 9:32, she reposted a message that an evacuation order had been issued for Alexis Creek and Hanceville. Twenty minutes later, she asked if there was anyone who could pull a nightshift, help unloading horses and packing water. "Everyone here is getting tired," she explained. "Due to the new order, it's probably going to get busy." At midnight, she reported that she was still able to take horses, but also pointed out, "If people are able to travel to Prince George, the air quality is better and the horse community is in full swing and ready for you."

By July 9, Lana was not only accepting horses at the stampede grounds but also overseeing their move to a safer place. For most horses, that was Prince George. As animals came in, Lana wrote down their names and ages in a simple lined notebook, described them, jotted down if they had brands, recorded behaviour ("halter broke, loads good, hasn't been tied yet"), dietary preferences ("grass hay for these horses, no alfalfa/grain"), medical needs ("on Pergolide[36]—syringe with pill and water in mouth") and the names and phone numbers of the owners.

Her friend Chelsea sent her a list of the licence plate numbers of the trailers she was sending down. When volunteers came to the grounds with trailers to take the horses out, she would record where the horses were headed and help to load them. "They'd go out, the next trailer would pull in and we'd load, and they'd go out. It was just like an assembly line and it went like that for twenty-four hours. Horses were coming in as fast as we could ship them out, pretty much. We had a little system," she said, modestly. I thought it was a marvel of organization, both flexible and thorough. When the rescue was over, no horses were lost and none lingered behind unclaimed.

Lana also retrieved horses left to fend for themselves. People who had no means of getting their animals to safety would frequently free them and hope they could find their own way. Volunteers working with Lana sometimes came across horses wandering along the roads and picked them up. Then Lana would post a description on Facebook:

July 9
7:52 AM

2 horses found on the highway loose are now safe at the Stampede Grounds. A sorrel gelding with a star, French shoes and a brand and a sorrel gelding with a partial left hind sock and blaze. Found by Sugar Cane.

Everyone was sharing Facebook messages so liberally that it didn't take more than four or five hours for the owner of these particular "orphan" horses to send Lana a message. She was so pleased someone had found them and was looking after them. Lana then arranged for the horses to go to Prince George. "[The owner] contacted me afterwards and was just so thankful," Lana remembers, observing that the fires brought people together who would probably otherwise never have known one another. Many people came to her and expressed their gratitude for what she had done. "I think it changes a lot of people for the better," she said.

When Fox Mountain was evacuated in the afternoon of July 9, a number of people who lived there were working in Williams Lake and weren't allowed to go home. Lana got permits from the RCMP and the CRD to send a few people who knew the area very well into their properties to find and retrieve the animals. When I asked Lana if she started getting calls from people whose horses were stuck behind the lines, she said, "You bet. And my phone blew up. It was crazy.

"And then I had CBC News trying to call me," she remembered. "I actually felt bad because one lady said, 'You know, I really want to do a FaceTime interview with you. I want to talk to you about what's going on there right now.' It was so chaotic, I had to say to her, 'I'm so sorry. I just do not have the time to do that with you.' I literally did not have two seconds."

By mid-afternoon, it seemed Lana's commodious ark was full:

2:47 PM

Anyone who has a load of horses NEEDS to head to Prince George. DO NOT come to the Stampede Grounds. We do not have room.

Lana got messages about other options and reposted those. The Nechako Valley Exhibition Society in Vanderhoof had fifty box stalls, twelve large pens and a livestock pen available. Lise in Cranbrook had pasture for twenty or so horses, as well as two stalls and two paddocks.

An hour later, Lana posted a handwritten note from Leo on Fox Mountain, who had three horses that needed to come out. He included an address and a phone number. Lana added:

3:47 PM

Can anyone help this fellow?

Evidently someone could, because later she posted:

Horses are safe—thanks everyone!

By early evening, Lana needed to clarify that she was offering rides to Prince George for horses that were currently on the grounds.

7:16 PM

No we are NOT on fire. We are just taking precautions. Some people including myself have sent horses to Prince George where I know they are being looked after.

At 7:51, a mule and its owners needed help. The couple were on Old Soda Creek Road, walking the mule out of their place. And then the update: the mule had arrived at the stampede grounds and was fine. Just two minutes later, Lana posted that she was once again able to take more horses.

7:53 PM

We were able to move horses to Prince George so we have more room here.

At 8:22, Lana's phone was dead from "trying to keep up with all the posts and questions." And then a comment by Froukje appeared on Lana's Facebook feed. She was out of town and needed her horse moved out. She was in and out of cell range. Kylie commented that she had trailer space for six to go from Williams Lake to Prince George. Froukje's horse made it to the stampede grounds and then she started looking to get her to Prince George:

> If you can, could you take my horse? Her name is Maple and she's a five-year-old mare standing at 14.3. She is chestnut with a white dot on her nose.

On July 10 at 10:32 a.m., Lana reported that she had space at the grounds if people wanted their horses to go to Prince George. Twelve hours later:

> July 10
> 10:30 PM
>
> We have horse trailers en route. You will have to stay with your horse and help load and give the driver a piece of paper with all the information on the horse/horses. The horses cannot just be dropped off anymore as we are not going to be able to continue their care.

Many people helped to relocate the horses. Logan Piesse, a twenty-six-year-old trucker from Alberta, made three trips between Prince George and Williams Lake on July 10 alone. That day he drove fourteen hundred kilometres with his six-horse trailer, going from morning until night, through the thickening smoke, eating hardly anything. Lana called him "a gift." None of the volunteers were paid for their time, although some companies donated gas cards so at least they didn't have to pay for their fuel.

On July 11 at 7:33 a.m., Lana reported that as far as she knew anyone coming from the north to take horses away was considered an essential service and would not need a permit to pass the checkpoints. But she cautioned that she did not know if or when this could change.

Horses had arrived at 2:30 in the morning, she said, and she had no way of knowing how many more would come.

At 8:54 a.m., she related that more horse trailers were coming and would be allowed to pass the checkpoints. She advised people who were bringing horses destined for Prince George to bring a note for the driver. This should explain how they could be contacted and give information about their horses. She promised that horses would be fed and watered on the grounds as long as possible, but also pointed out that her volunteers had to leave to look after their own families and animals. She herself was driving up to Prince George to retrieve her truck and horse trailer so that she could assist with the final evacuation effort.

And then it was over:

July 11
7:37 PM

They are not needing any more horses hauled out of Williams Lake. Thank you so much to everyone who helped either in a big or little way. You have no idea how much you are appreciated and loved. Please share so no more trucks and trailers head down to us.

Lana was going to stay in Prince George. She has restricted lung capacity because part of one was removed a few years ago, due to cancer. The smoke blanketing Williams Lake made breathing too difficult. When she left, only half a dozen horses remained. They belonged to a few people who were staying at the adjacent campgrounds and taking responsibility for trailering their animals out if an evacuation order came.

I was curious about how Lana had managed with so many animals unfamiliar with each other. I had understood that horses didn't like to be with strange horses. "I've raised horses for twenty-five years and been involved in the horse industry all my life," she said. "Horses will get across the fence from each other and they'll fight. They'll be silly and they don't like each other. But the strangest part was that we never had a problem with any of those horses. Everybody was put into a pen and they didn't know each other but they didn't do anything. Horses that had never been hauled before in their lives got thrown into trailers and taken. It's like they *knew*. It was the oddest thing. A lot of these horses had never been to town before. We found a couple of them running down the highway. Somebody had let them loose and we caught them and threw them in a trailer and brought them in. I honestly think they just knew they were going to be safe. Everybody was calm. It was very quiet. There was no nickering at night or squealing. It was the same thing when they all got to Prince George. I had a four-year-old who had never been in a stall in her life. And, you know, I hauled my horses up there and threw her in a stall and said, 'Okay, this is where you have to live.' It was just like she had done it all her life."

"They're just like people. *They* behaved better than normal too," I said. We both laughed. I asked Lana how many horses she and the other people helping her had saved. It was hard to estimate, she told me. She thought that around three hundred was the peak number of horses at the grounds. But while that number may have been constant for a while, the individual horses were always changing. As some were driven away, others came in to replace them. For their efforts, in 2018 the Horse Council of BC gave Lana and her friend Chelsea Volunteer of the Year Awards.

I thought it was a remarkable that the rescue of so many animals took place without anyone being *ordered* to do anything. Like musicians in a band, people were eager to contribute; they just needed someone to coordinate their efforts. The result was an astonishing crescendo of goodwill.

CHAPTER 11

EVACUATE, EVACUATE!

"THE FIRE CENTRE IS EVACUATING ..." The Cariboo Fire Centre's radio dispatcher broadcast the news at 3:00 p.m. on July 7. Trevor Briggs and his Blackwater Unit of firefighters based in Quesnel heard the report as they were on their way to Williams Lake to help the local fire department. Coming south on Highway 97 had been apocalyptic. Trevor wrote, "After driving by the Green Mountain fire, we saw multiple lightning strikes—all of which were instantly turning into smoke columns due to severe fire weather, some of which merged together as we drove by. Nearing McLeese Lake, halfway to our destination, we could see clearly out west toward Nazko and the Chilcotin, and it was just a scene of massive smoke column after massive smoke column. At this point, we knew it was going to be a long summer and a crazy few days ahead for us and the affected communities."[37]

Fire and ice. That's the story of lightning, according to "fulminologists," who study the subject. Collisions between ice crystals in the middle of a storm cloud create friction, which in turn produces electrical charges. Lighter particles rise to the top of the cloud and acquire a positive charge while heavier ones sink to the bottom and become negatively charged. The electrical charges keep building until a giant spark—lightning—occurs. The amount of energy released from the rubbing of these ice crystals is astonishing. In a fraction of a second, lightning heats the air around it to 30,000°C—five times hotter than the surface of the sun. When it hits a tree, it can cause the water in the cells beneath the bark to boil instantly and explode. On July 7, thousands of lightning bolts struck the Cariboo and Chilcotin, discharging billions of joules of energy. No wonder flames erupted all over in the tinder-dry forests and grasslands. (It was so arid that some cowboys in the Chilcotin removed their horses' shoes, for fear they would strike a rock and emit sparks.)

Several fires were burning in and around Williams Lake when Trevor arrived: at Spokin Lake to the east, at 150 Mile House to the south and in the north near the community of Wildwood and Williams Lake Airport. A pine-beetle-ravaged forest around the airport was ablaze and threatening the Cariboo Fire Centre. This was a $7 million building[38] just commissioned in March 2017. Jessica Mack, a spokesperson for the centre, told me in an email that twenty-five staff members, whose jurisdiction covered 10.3 million hectares,[39] had all been ordered to leave. Also at risk on the tarmac and waiting to be deployed were six helicopters out of thirty that the centre had hired for operations, as well as a "bird dog" for air traffic control and two air tankers for dropping fire retardant. Bird dog planes are used to guide the flight of the water bombers and direct them where to drop their payloads, performing an essential coordination function, like the air traffic controllers at an airport.

Trevor and his unit were first sent to work around Wildwood. In total, twenty-seven firefighters were on the ground either there or at the airport. In the evening, Trevor and his group were called to conduct a controlled burn around the airport. Conditions were good; the winds had calmed down. While an operator in a dozer was creating a fireguard, Trevor helped by conducting planned ignitions. He worked until the early morning, when half of the crew at the airport, including him, were allowed to catch a few hours of sleep. The other half continued to extend the fireguard.

When day dawned, Trevor and the firefighters who had been allowed to rest returned to the airport to relieve their colleagues. "The guard line was nicely done," Trevor wrote. But there was a problem with the berm, a ridge of dirt built to help control the spread of rolling embers. On one side of the firebreak was green, unburned land; on the other, burned or partially burned ground. The berm was on the scorched side instead of on the green side.

After that was fixed, the crew increased the size of the buffer by spraying the green land with water and by burning off the remaining fuel on the other side. These operations were successful enough to allow the staff at the fire centre to return to work. The fire overran a section of runway at the northern end of the airport and two hangars were destroyed, but none of the aircraft were damaged—they survived to fight another day.

BUT THE SUGAR CANE FIRE to the southeast of Williams Lake and the Wildwood fire to the northeast were still burning. They were approaching each other in a pincer-like action. At 2:00 p.m. on July 9, an evacuation order was issued for Fox Mountain and the Soda Creek area. The Sugar Cane fire was just two kilometres away from Brian McNaughton's house. He lives on Fox Mountain about ten minutes north of the city and is the general manager of the Federation of BC Woodlot Associations. Knowledgeable about forests and fires, he was vividly aware of the hazards near his own house. He told me, "Across the street, there was a lot of dead material. The understorey had been killed by spruce budworm seven or eight years before. For some reason we have copious amounts of green lichen. So we had the makings of ladder fuels and a holocaust." Brian and his wife left their house and went to stay at his sister's place in Williams Lake.

The Wildwood and Sugar Cane fires merged. That blaze and another one, the White Lake fire to the northwest of Williams Lake, were growing aggressively. On July 10 at 6:22 p.m., a city-wide evacuation alert was issued for Williams Lake. Walt Cobb, serving his ninth year as mayor, was monitoring the situation closely. To add to his difficulties, Williams Lake had more people in town than it normally would. Two evacuation centres, one set up by the Cariboo Regional District and one organized by the town itself, had welcomed evacuees from 100 Mile and from the fires to the west in the Chilcotin. Fortunately, the horses being sheltered on the stampede grounds were already being moved out.

To prepare for the possibility that everyone might have to leave Williams Lake, Walt said, "We evacuated the elderly and anybody with respiratory issues." The hospital took its patients out, most of

Superintendent Michel Legault evacuated his officers when the log yards in Williams Lake were at risk of catching fire.

them to Prince George. And the RCMP sent 175 more officers to Williams Lake in case the extra police presence was needed. Walt recalled, "We had the community divided into twelve different sections so that if in fact we needed to evacuate we wouldn't create a traffic jam. We had muster points for people who couldn't drive or didn't have vehicles. The plan was for everyone to go north on Highway 97 to Prince George."

By July 15, the White Lake fire to the northwest of Williams Lake had burgeoned to thirty-eight hundred hectares. About 3:30 in the afternoon, Walt got a message from the fire centre to be prepared because the fire had jumped the Fraser River at the Rudy Johnson Bridge. He recalled, "We watched it for a bit and then it started moving toward Williams Lake at about forty klicks. So it was coming fast. We got the word: 'Evacuate, evacuate.'" At 4:59 p.m., Walt signed the papers ordering an evacuation of the town.[40] "We phoned the buses, told them to go to the muster stations, phoned the RCMP, told them to get everything lined up, and start getting ready to direct traffic."

Superintendent Michel Legault was doing a tour in Williams Lake as an RCMP Bronze Commander. Later I spoke to him in Chilliwack, where he was the Officer in Charge, Pacific Region Training Centre. A brawny, dark-haired man with a moustache and a Québécois accent, he said, "The first part of the evacuation went really well. The challenge was we had a planned escape route north, and when the evacuation order was given, the north highway was cut off by fire. So now we had to switch all of our thought processes." Evacuees had to go south to 100 Mile House, then east along Highway 24 to Little Fort, and then south again on Highway 5, where they could register at the Sandman evacuation centre in Kamloops. But this meant the evacuees would have to pass through the evacuation zone around 100 Mile House. Michel said, "We were putting people in an area where we were not supposed to send them, in a sense, so we needed to make sure that all of our area that was under order was well guarded. We didn't want people to say, 'I'm just going to turn here, park, wait this out.' We couldn't do that. So we deployed along the south highway. We monitored the traffic and we had people on Highway 24. We had people all the way down, midway to Highway 5, because we needed to tell them to go south to Kamloops, not north to Prince George."

Chief Superintendent Dave Attfield was Gold Commander of the RCMP policing efforts during the fires.

At 5:30 p.m., Michel called Staff Sergeant Svend Nielsen at the detachment in 100 Mile House and said, "Williams Lake is being evacuated."

"Which way are they going?" Svend asked.

"They're going to be in your area in forty-five minutes."

"All righty then."

Svend told his officers to drop everything they were doing. "We have forty-five minutes to close down the area and make sure no one stops anywhere." Svend was able to get some help. Michel had asked his superior, Chief Superintendent Dave Attfield, for assistance and he had been able to secure the K Division Tactical Unit from Alberta with forty officers and a dozen vehicles. These officers, who are specially trained to deal with a variety of emergency situations, had arrived in the region on July 13 and were able to help patrol the exit route.

THE EVACUATION OF WILLIAMS LAKE had ripple effects across the Interior of BC. In the tiny hamlet of Little Fort, Pam Jim, who together with her husband, Kam, owns Jim's Food Market, was also notified about what was happening. "We got a call from emergency services telling us that Williams Lake was being evacuated and asking us if we could remain open," she said. "We made the decision to stay open twenty-four hours because there were five thousand people coming down."

Brian McNaughton and his wife, who had previously gone to Williams Lake for safety, were evacuated for a second time. "We were originally told the highway was only going to open north so I made some contact with friends up north and said, 'We'll see you in a few hours,' and then we were told that they couldn't open that route. We had to go south. So we cancelled those plans and I called my niece and said, 'Guess what? I'm coming for a visit—and bringing some in-laws.'" In two cars, Brian and his wife, together with his sister and her husband, followed the long, slow queue going out of Williams Lake. They were headed for Kamloops.

At 6:15 p.m., Svend (wearing what he called his "ritual shorts and t-shirt") and the officer in charge of the tactical unit from Alberta stopped the first vehicle that approached the detachment office. They explained to the driver that he had to wait a few minutes before proceeding along Highway 97 because a tactical team was coming to make sure that everyone got safely through to Highway 5 and out. Moments later the team arrived, got ahead of the evacuees in their vehicles and soon that lengthy line began snaking forward again.

Back in Williams Lake, Michel got a call from the BC Wildfire Service at 7:30, telling him that it was now safe for vehicles to go north. The window lasted about two hours and it allowed the RCMP to divert traffic to Prince George, taking some of the pressure off the southern escape route. At 9:30, Michel, who had more or less been pinned to his desk in the command centre in the basement of the Williams Lake City Hall since the evacuation order had been issued, decided he needed a breath of fresh air. This would be metaphorical fresh air, since it was still smoky outside. When he emerged into the parking lot and looked north, the deep red glow in the sky disturbed him.

There were several sawmills at the northern edge of Williams Lake, and Michel had heard worst-case scenarios. If the log piles caught, a firestorm that generated its own wind could pick up embers and burning debris and blow them onto the town. If that happened, no one could stop the inferno. Williams Lake could not be saved. The two largest mills and those of most concern were West Fraser Timber and Tolko Industries. Lying between them and the White Lake fire were three breaks: the first guard was one kilometre north of the mills, the second was at the three-kilometre mark and the third at the five-kilometre mark. Just over a kilometre south of a smaller mill was Columeetza Secondary School, which the RCMP was using to house the 175 officers who had come from out of town to work in Williams Lake. Since the wind was from the northwest, this put the school in the line of fire. Michel sent a couple of officers to look at the mills and see how many logs were stacked up in the yards. He wanted to know how much of a risk he was facing. "You've got about six thousand logs," they reported back. *That's not possible*, he thought. Having seen the mills, he was expecting a much larger number. He sent his officers to look again. But this time he asked a younger staff member, who was from Williams Lake and familiar with the local geography, to go along. The second report was more in line with what Michel was thinking: sixty-six thousand logs. There were thirty thousand at each of the two largest mills and six thousand at a smaller one. He also learned that thirty fire engines were hosing down the logs. But instead of alleviating his anxiety, this information only emphasized how serious the danger was. (Later, when I visited Williams Lake to do some research, I was stunned by the size of the log piles at the north end of town. I had never seen so much cut timber in my life. Some of the piles were ten metres high, twelve rows deep and a city block long. I could see why Michel would have been concerned.)

Michel phoned Glen Burgess, the fire official I'd met in Kamloops. He did tours as an incident commander in several places and at this point was in Williams Lake.

"Where are all your firefighters sleeping?" he asked.

"They're at the airport."

"Why are they at the airport?"

"Because it's the safest place we're going to have right now. We've burnt around the airport."

At 10:00 p.m., Michel decided to move his officers to the airport too—their cots, their clothes, everything. "I did have a conversation with my command group at that time and I said, 'This is not going to be a consensus-concept decision. I'm not taking a chance. I'm just telling you we're going to move forward, unless you have a major objection or a different scenario where you want us to move somewhere else.' Everybody was like, 'No, we're all in. That's the right thing to do.'" A couple officers from the logistics team went up ahead and assessed the situation. One of them told Michel he had space for sixty officers in the terminal.

Michel asked, "What else is there?"

"We have hangars, but there are planes in them."

"Well you're moving planes, so cut the locks, get in, open up so that we have a space. We'll pay the repairs, everything that needs to be done after the fact, but right now I need to put people there so they can sleep because they need to go back to work tomorrow."

Michel told his command group that he would be staying the night at City Hall. He also said, "You know, I'm not asking you to stay. I'm okay with you going." All four said, "No, no, we're okay." They remained behind while everyone else moved. Michel explained, "Where we were was the best place for us to monitor, to get information. I had already asked our technical group to move half of our command centre to the airport so that if we were to go there, it was ready for us. But it wasn't set up. If I was going there right away, I had very limited communications, very limited access to everything that was required for us to make proper decisions." He added, "We had set ourselves up a route and we knew that the vehicles were there ready to go if it really hit the fan." He also continued to track what the fire was doing.

However, as Superintendent Dave Attfield reminded me, "In this case, you had dynamic fire behaviour, which was changeable and unpredictable." In other words, it was challenging to say what would happen. I met Dave at the BC RCMP Headquarters in Surrey. He was the Gold Commander during the fires, responsible for all areas in BC in which major fires had erupted. His regular post is Chief Deputy Criminal Operations Officer for Core Policing. Superintendent John Brewer, who was a Silver Commander during the fires, was also at the meeting.

John saw Michel's decision this way: "He was acting in the best tradition of the RCMP, saying 'I may die tonight but I'll make sure that my members get out.'" John is a member of the Lower Similkameen Indian Band and is normally responsible for a number of portfolios, including First Nations Policing. He said, "Thinking about that still gives me chills. He slept right at City Hall knowing that was ground zero if it ever went." John, who served in Afghanistan in 2010 as a NATO Senior Police Advisor, added, "Outside of Afghanistan, that evacuation was one of the hardest nights I had ever had. We stayed up all night here too."

THE STEADY STREAM OF TRAFFIC out of Williams Lake continued. As night fell, people on the road could see a string of lights ahead of them and a dark red glow behind. They had no idea whether they would see their homes again. Ten thousand people left the city in seven thousand vehicles. The convoy took nine and a half hours to pass through 100 Mile House. It was a gear-laden exodus. Brian McNaughton said, "We'd recognize a husband with a pickup pulling a fifth wheel and the wife in the next pickup pulling the horse trailer and the oldest child driving the third vehicle." Svend told me that an RCMP officer from Port Moody, who was off-duty, filmed the whole procession—"every single car." Later Svend watched the first four or five minutes of the video. "But I stopped. It just got to be too much."

The evacuation went surprisingly smoothly, especially considering the number of vehicles on the road. "People were extremely polite," Brian said. "I didn't even see a hint of road rage. We didn't pass anybody and nobody passed us all the way from Williams Lake

Superintendent John Brewer, who served in Afghanistan, remembers the night of July 15 as one of the most difficult in his life.

right through to Little Fort." One woman fell asleep at the wheel and slid off the highway. But she was not seriously injured.

At Little Fort, the RCMP was stopping traffic from going west on Highway 24 so the evacuees coming east could occupy both lanes. Pam Jim told me that, despite this arrangement, the vehicles were bumper-to-bumper, towing trailers on the steep hill leading down from McDonald Summit into the village: "People were hauling loads of quads. There was a snowmobile in a twelve-foot boat on a trailer. Chickens were in a house-type cage on a Suburban."

Jim's Food Market was founded by Kam Jim's grandfather in 1919. A Cariboo institution, it was close to celebrating its one hundredth anniversary. The Jims had deep roots in their small but historic community, which got its start in the 1840s as a stopping point on the Hudson's Bay Company Brigade Trail. As Pam watched the train of vehicles passing by, she began to think about the people in them. They were under stress and didn't always know where they were headed. Many had been evacuated just before suppertime and didn't get any food. They had been on the road for seven hours and were tired and hungry. Pam told me, "Jenny and Wyatt [her daughter and her boyfriend] and I started making some sandwiches for them. Some of them just about cried when they got that sandwich. They'd ask, 'Do I have to pay?' and I said, 'No you don't. Take a sandwich.' They didn't even ask what was on them. We started running out of bread. So we made wraps. Then we ran out of energy and food about 2:30 or 3:00 a.m. We gave away a couple hundred easy."

Pam closed her kitchen, but she wasn't able to go to bed just yet. The gas station was almost out of fuel. When she called Husky to deliver more, she was told no drivers were available because they had all worked the maximum amount of overtime permitted. She called the RCMP, who called Husky, who then agreed to send a delivery from Kamloops. It arrived at 4 a.m., just before the pumps ran dry. "I gave the driver coffee, whatever he wanted," Pam said. "I was so appreciative of him." And then she finally turned in.

BY 3:00 A.M., THE WINDS had calmed down. As Michel recalled, "We were told that the fire had basically changed direction at one of the breaks. If my memory serves me well, I think it was the three-kilometre break. It actually stopped. But it also had crossed Highway 97, which cut our highway access to the north again." Still, the situation seemed more stable, so Michel decided to pay a visit to the Williams Lake Airport. It was like a refugee camp: "I had people sleeping underneath the wings of planes on the tarmac. I had some people on couches, some people in chairs and some people on the floor. People pretty much everywhere."

Half an hour later, when he was satisfied that his members were all okay, he returned to the command centre. At least the immediate threat to Williams Lake had gone down a notch. Michel told the officers at the command post that they could catch some sleep—but not a lot of sleep, mind you. Michel was back on his computer at 5:00 a.m., updating his superiors on the latest developments.

At 7:00, Michel did his normal morning briefing with the people coming off shift and those who were coming on. "I updated them on the evacuation and told them what an awesome job they had done. I told them very clearly, 'You know, my goal over the next twenty-four hours is basically to get you shelter, food, a place to have a shower and a place to do your normal daily functions.'"

However, it wasn't as easy as it might sound. "We had a moment of reckoning," Dave Attfield told me. The RCMP had been relying on the Cariboo Bethel Church to provide three meals a day for about three hundred officers. A dozen volunteers had been cooking up a storm, starting with a dinner on July 12. (They had help. The pastor, Jeremy Vogt, told me that the Cariboo Community Church, a sister organization, made bag lunches. Local grocery stores, Safeway, Save-On Foods and Wholesale Club donated food or sold it to the church at wholesale prices. People in the community brought baking. One woman brought six pecan pies.) On July 15, volunteers at the church served breakfast and lunch and were just about to dish up a Mexican-style enchilada casserole when they were evacuated along with everyone else. They scrambled to put things away and left as quickly as they could.

The RCMP had some military-style ready-to-eat meals, but the supplies would only last until July 17. And on that day, they would be marginal: "We had a breakfast and half a lunch for all our members," John recalled. Dave wrote in his briefing notes to his superiors in Ottawa, "We have lost the ability to feed our members." He immediately received an emergency procurement contract worth about $1 million. This would allow him to establish a field kitchen that could supply meals and boxed lunches for thirty days. It would also let him purchase more military rations to use until the kitchen was running.

It took several phone calls to locate a supplier who had a stock of military rations on hand, but eventually one was found—in the US. "They were palletizing them and putting them on a plane out of Chicago," John said. "It was landing at six o'clock in the morning at Vancouver airport. So I had two members there in vehicles as soon as that plane landed. 'You get out there and start loading,' I said." When one of the young corporals that John charged with the duty got to the airport, he discovered a hitch. "He phoned me," John recalled. "He said, 'Sir, the delivery has to go through customs. It has to be inspected. We can't do anything. It's going to take almost a day,' I'm like, 'I don't care how you get it. At gun point,' I said. 'I'm good with that.' It was a verbal stare-down with customs to get it through because nobody understood just how close we were to running out of food. Even though there's a grocery in Williams Lake, it had been evacuated. You'd have to have pretty exceptional circumstances for the RCMP to kick in a door and start looting for food. It worked out pretty well for us. We made it through, but we were a meal and a half away from not being able to feed, it was that close."

Top: Conair's Air Tractor 802 Fire Boss planes in action at 108 Mile, July 6.
Bottom: Smoke roils up northwest of Little Fort, July 7.

Top: Fire threatens the Bonaparte First Nation, July 7.
Bottom: Bomber flies toward Williams Lake, July 7.

Lee Todd reconnoitres over Riske Creek, July 7.

This chopper was a sitting duck when an inferno hit the tarmac at Williams Lake Airport on July 7.

The Elephant Hill fire melted this trailer into a Daliesque sculpture.

Smoke obscures the sunrise at Sheridan Lake, July 8.

Top: A couple of signs are all that remain of the historic café and gas station at Lee's corner.
Bottom: Roland and Udette Class stand in front of skeletal trunks near Hanceville.

Top: Chilcotin Towing became an icon of destruction on Highway 20 in Riske Creek.
Bottom: Trees are wreathed in smoke near Hanceville.

Top: Fire creates an orange panorama behind the storied Becher Ranch in Riske Creek on July 15.
Opposite: Fire retardant is deployed near 16 Mile House on July 13. Altogether,
12 million litres were used across BC in 2017.
Above: A lurid sky looms over the gas station at 70 Mile House, July 14.

Top: Fill 'er up! Canadian Forces roll into 70 Mile House, August 7.

Bottom: Firefighters build fireguards at Slater Mountain on July 17 to protect Williams Lake.

Top: An early morning safety meeting convenes on Slater Mountain, July 22.
Bottom: Fire leaping across Highway 97 on August 1 catches Delta firefighters off guard.

Top: A Mexican CONAFOR firefighter poses with George Keener's donkey on Slater Mountain, August 6.
Bottom: Trees candle after firefighters light a prescribed burn on Slater Mountain, August 4th.

Top: Conair's Fire Boss 88 lobs water on the Lamb Creek fire, south of Cranbrook, August 10.
Bottom: Firefighters observe an imposing blaze west of Quesnel, near Nazko, August 12.

Top: Quebec bombers skim across Green Lake, August 16.

Bottom: Conair's Fire Boss 73 and 87 work on dousing the Lamb Creek fire, August 29.

Top: Dark smoke wreathes Green Lake on August 30.
Bottom: The wind shifts on September 2, sending the Elephant Hill fire south.

Top: One year after the fire went through, trees east of Riske Creek are a testament to what happened.
Bottom: Paradise regained: fireweed flourishes the summer after fires created this clearing beside Sheridan West Forest Road.

The winds died down and the threat to Williams Lake abated. Michel arranged for two local hotels to accommodate his officers. Of course, the hotel staff had been evacuated. "We became innkeepers," Michel said. A mobile kitchen in Alberta's oil patch was shipped to Williams Lake. Food was procured as well and on July 20, the officers were finally able to enjoy a hot, freshly cooked meal. Michel left Williams Lake the next day. His first week-long tour on the fires was over. He came back, however, after the evacuation order for Williams Lake was downgraded to an alert on July 27 and the community was welcoming its residents back. "That was a good feeling," he said.

BRIAN MCNAUGHTON AND HIS WIFE first heard about the evacuation order for Williams Lake being rescinded on social media, while they were staying at his niece's place in Kamloops. "My wife said, 'We'll just kind of tidy up and we'll leave tomorrow morning,'" Brian recalled. "I said, 'We're out of here.'" They packed up and a few hours later were home again: "Turning the corner and then turning in the driveway and seeing the house still standing was surreal. I had in my mind I was going to come back to brown lawn and everything shrivelled. I had two tomato plants that didn't get watered for three weeks and survived. My wife had plants all over the deck and one plant died but the lion's share survived. Still, the nerve ends are raw. If you watch a lot of catastrophes in the world—you know, hurricanes in Florida—you go, 'Oh those poor people.' But now I watch the people in Burns Lake and I know exactly what they're going through and it sends shivers down your spine. You look around and you go, 'It could happen to me again.' It's a very raw emotion. I don't think there's anyone that goes through this who doesn't experience lingering effects."

CHAPTER 12

GO SAM, GO

"I NEED TO KNOW WHO'S willing to fight fire," Gord said. "You know, it's not *if* we get a call, it's *when* we get a call." Samantha Smolen, or Sam, as she is usually called, didn't hesitate at all. "I'm super keen," she said. For five years, Sam had worked as a forest technician for Gord Chipman of Alkali Resource Management. She had surveyed forests, planned tree harvesting, supervised planting and managed cone picking. Though she had never fought fires before, she had taken her S-100 training, a basic course in fire suppression and safety. She was eager to apply what she had learned and at age thirty-two was ready for some adventure.

Sam told me about her July 7 meeting with Gord as we spoke in a small café in North Vancouver. A blond woman with a warm smile, she had gone to school at Handsworth, a high school my kids had also attended, and graduated from UBC with a B.Sc. in Natural Resource Conservation. She was taking additional courses there to qualify as a registered professional forester.

In early July, Gord already had the usual number of crews available on standby to fight fires. But because the summer of 2017 had all the signs of being exceptionally hot and dry, he was pretty sure BC Wildfire would want to hire more than the regular workers. Normal logging operations were shut down due to the hazardous conditions, so he called a meeting of his dozen forest technicians in Alkali Lake to see if any of them might be interested in helping out on the fires.

Rookie firefighter Samantha Smolen took on spark duty early in the morning of July 16 on Slater Mountain.

There was only one difficulty for Sam. She had arranged to visit Vancouver for a few days, so she couldn't begin right away. Gord agreed to let her start after her trip. But once she got down to the coast, she realized fires were erupting all over the province. "I was visiting family," Sam recalled, "and I was worried about my friends in Alkali. I was starting to freak out because I felt like I needed to get back there to help. So my roommate said, 'Well if you want to go, I'll drive you back.'" Generally the trip would take about seven hours, but by July 8, several highways leading to the Cariboo-Chilcotin were closed due to the fires: Highway 97, Highway 97c, Highway 1 and Highway 99. Still, in her years up north, Sam had become acquainted with the *ways*, the network of dirt roads of varying degrees of drivability that criss-cross the backcountry. She and her friend got through, but it took them twelve hours.

Sam arrived in Williams Lake late on Sunday, July 9, but she went to see her office administrator anyway. She got her Nomex, the *de rigueur* clothing for firefighters, and was assigned to a crew. Glen Chelsea was the crew boss; Dave Dan, Jason Walch and Brett Harry were the other members of the team. Sam was the only woman in the group and the only non-Indigenous person. Being in a minority position like that didn't bother her one bit. She praised the fellows she worked with: "They were great guys." However, she did recall one of the problems of being a woman in a male environment: "It was pretty funny at first being the only girl. I was laughing 'cause we have to roll up the hoses a certain way and I could barely bend in these pants." (Eventually, she got a pair of Nomex more suited to a female form.) The guys teased her, but she got their respect for her skills with a compass and GPS points, which she had acquired as a forest technician. When Gord wanted her to return to regular duties after her first fourteen-day tour on the fires, Glen told her he had complained: "Gord, no, you're not taking her. If you take her, I'm leaving." Gord let her do a second tour with Glen.

On July 10, a co-worker took Sam and the rest of her crew to Riske Creek, a small ranching and logging community in the rolling grasslands fifty kilometres west of Williams Lake, along Highway 20. There they met up with Gord, who had brought five more crews with him. For three days, until the BC Wildfire Service could get one of its incident commanders on-site, he also helped coordinate the activities of the logging companies supplying heavy equipment.

Riske Creek dates back to the 1860s, when L.W. Riske, a Polish immigrant, established a flour mill and sawmill in the area. The community sprawls; houses are a good distance apart from one another. But it is also tight-knit; many of the ninety residents know each other very well. One local rancher joked with me that as he had lived in Riske Creek for a mere thirty years, he was still considered a newcomer.

The Alkali Unit Crew—Jason Walch, Dave Dan, Brett Harry, Samantha Smolen and Glen Chelsea—helped to save Williams Lake.

From July 7 on, when multiple lightning strikes hit the Chilcotin, the locals had been active in their own defense. Working long, gruelling days, they had constructed ten kilometres of firebreaks. "It was amazing how much guard they built just trying to get in front of the fire, trying to cap it off," Gord said to me. "But the fire kept jumping these guards, no matter how many they put down. It was so dry the fire would burn right across the guards. There was so much wind. The embers were the biggest issue. They just kept on blowing across the guards and then started another fire."

"At first, things were disorganized," Sam said. "The government ministry guys, the red shirts, weren't there yet. We didn't do a whole lot that first day. Gord went up a few times in the helicopter to check things out. I was a little bit nervous because I had never done it before. But I felt prepared." Once they got their bearings, Sam and her teammates were able to make headway. Using hoses and hand tools, they were "blacklining"—pushing the fires back about thirty metres from roadways and making wider guards. But on July 15, the winds picked up again. They were working on the road leading to the Riske Creek transfer station when the fire jumped the buffer zone the crew had created, as well as the road itself. Sam had never seen anything like it. She was astonished by how easily the trees caught fire and then crashed around them, scattering sparks as they fell. The crew retreated to Highway 20, and then the members were radioed instructions to go to a house in danger of succumbing.

When she came to the white house with blue trim in a grove of trees, Sam recalled, "You could just see the flames. Your adrenaline is pumping and you want to stop the fire from burning down this house. I showed up and someone said, 'Are you doing anything?' And I said, 'No.' He said, 'Grab this hose.' So I just grabbed it and everyone around was kind of busy or frantic or whatever and I crawled under a fence and started putting out the fire. No one was helping me. It was one-and-a-half-inch hose. We were taking turns rushing in around the house. It was so smoky and the guy was bombing water 'round the house with the helicopter. We had danger tree fallers as well, who were falling some trees and cutting them into smaller pieces to mitigate the spread."

Blake Chipman was on the water tender that was supplying Sam's hose. He was Gord Chipman's father, and had thirty years of experience in the woods, as a logger and firefighter. The BC Wildfire Service had hired him in the springtime. "We're usually on standby from morning 'til night. When a fire happens, they call us and we go," he explained in a phone interview. He remembered Sam vividly. "I'm on top of my truck with a pump. And this is basically a big fire. When she went below the road, there was raging flames within fifteen to twenty feet of her. And she actually put that fire out. She just did her job. And if she wouldn't have got it there, then it would've continued on. A good, hard-working woman. Some of the other crews were quite disappointing actually. Fighting fire, you think everybody would try and, you know, do the best job they can. Yeah, but it wasn't the case sometimes. A lot of times it was just a lot of sitting around and people weren't really interested in working. It was hard to light a fire under them," he said and chuckled.

Sam said three contract crews from Alkali Lake and two from the Toosey First Nation in Riske Creek were deployed to save the house. Despite this effort, the fire was quickly becoming unmanageable. The winds were strong and the fire was "basically all around us," Sam said.

"That fire was actually casting the fire ahead of itself by about a hundred yards," Blake recalled. "It was throwing balls of fire the size of the hood of your vehicle. The trees were about 120 feet tall and the flames were thirty to forty feet higher than the trees.[41] My dog was in the truck and I had to roll up the windows so he wouldn't burn up because there was a lot of embers and shit flying around. Oh yeah, it was quite drastic there."

By then, Shelly Harnden had taken over as the incident commander for the Riske Creek fire. She decided the firefighters had better retreat and go back to the Old School in Riske Creek, an adult training centre, which the BC Wildfire Service was using as a base. "When we left that house," Sam said, "the fire was on both sides of the highway at one point." She was driving and as soon as she saw the flames ahead of her, she looked at Glen, her supervisor, questioningly. He said, "Just go, Sam, just go." She stayed in the middle of the road to keep the maximum distance between the vehicle and the fire. She wasn't too worried, "But you'd want to go fast," she pointed out. While she was waiting with the other firefighters at the base to receive directions, Sam learned that both Williams Lake and Alkali Lake were being evacuated and that it was now too dangerous for her crew to be in Riske Creek. "But there was no instruction like 'Come back at this time,' or, you know, 'Check in with this person or with me or whoever.' There was nothing. It was just like, 'I don't know. Just go.'

"We sat in the traffic trying to get into town while everyone was trying to get out. I don't know if 'scared' is necessarily the right word, but I was very anxious," Sam said. The steady stream of cars heading out of Williams Lake and the departure of thousands of people were strong signals that the town was not safe. "But I didn't know whether I was supposed to be leaving town with everyone or staying. I was just kind of following my gut and staying in town."

When Sam got back to her place, it was 9:00 or 9:30 p.m. Her roommates were gone and the families of her crew members had been evacuated. Sam said, "I asked my crew, 'Well, why don't you guys just stay with me? I've got beds and bedrooms.' So they stayed. Actually one of the guys put his gear in the wash."

"I texted Gord, my boss, saying, 'Hey Gord, I've got my crew with me at my place. Can you tell me what's going on?' I'm kind of stressed. I don't know what to do—if I should be leaving or if I can stay. Nobody told me that." Sam had got used to operating within a chain of command, being guided by clear instructions. But she didn't have any and she found the situation disturbing. "I texted Gord, saying, 'We're ready to go,' thinking he would give us some direction in the morning. But he texts me back, or calls me back, and says, 'Are you sure you're ready to go?' And I'm like, 'Yes, we're ready to go.' Then he asks, 'Can you be up at Slater Mountain in half an hour or an hour?'

"I told my crew boss and he's like, 'We'll be there.' He made my buddy take his gear out of the wash and put it on wet. When we drove up it was 11:00 at night." Slater Mountain is at the north end of Williams Lake, north of the sawmills. When sparks jumped the Fraser River, the mills were at high risk of catching fire and igniting Williams Lake. The Alkali crew's assignment was simple to describe: "Put out any fires caused by the sparks and prevent the flames from spreading to the mills." But it wasn't so easy to execute.

"We got up there," Sam recalled. "It was pitch black and we couldn't see anything except the red glow of the fire. The wind was howling. We could hear a machine bulldozing trees over to make a guard. Not even cutting them and throwing them aside or anything, just plowing them over. They had two guys on the radio. The line locator was out there on foot with a GPS, hanging boundary for the guy in the machine right behind him. We couldn't see a single thing except for the trees moving and falling down, the headlights of the machine and the glow of the fire. We could hear the wind roaring and they wanted us go out there and put out any sparks. I'm like, 'Oh my gosh, I don't want to get out of the vehicle. This is not safe.'"

LEE TODD, THE PRESIDENT OF Eldorado Log Hauling, is a friend of Gord Chipman's and together they had organized the operation. I talked to Lee at his company office on the lower slopes of Slater Mountain and learned that he'd had a hardscrabble start in life. After leaving his home in Riske Creek at fourteen, he'd hitchhiked

Lee Todd, the president of Eldorado Log Hauling, organized firefighting efforts on top of Slater Mountain.

to Vancouver, where he'd spent one winter living in a cardboard box. Eventually, he moved to Williams Lake, built a thriving business and appreciated the chances he'd had. "I love this town," he said. "It's vibrant, a very good town." He told me that in 2017, after working all summer on the fires, he was diagnosed with cancer and went through a gruelling eight months of chemo. Despite that, his zeal for the subject of fires was undiminished.

On July 15, Lee's wife phoned him to say that an A-Star helicopter had landed on their property, just a few kilometres away from his business, higher on the mountain. "So I go roaring up in my truck," Lee remembered. A BC Wildfire official emerged from the chopper and said, "Lee, run like hell, you've got twenty minutes and then you're toast."

"I'm thinking he's one of the higher echelons, he should know his stuff, so I get the wife down here and my motorhome and got my valuable belongings out of our house." But Lee could not resist having a look at the fire himself. He has been fighting fires ever since he was a teenager, so forty-six or forty-seven years. He jumped in the helicopter that he used for his business. "I went up and, excuse my language, but for fuck's sake! You know, when they'd seen it, it was probably a cat 5 or 6. I could see it from here—a mushroom atomic bomb. It was uncontrollable, with the wind in the afternoon. Only a fool would have been endangering people's lives by putting them on the smoke in front of that fire. I'm not that stupid. But as soon as I saw it, I thought, 'Man it's 4:00 or 5:00 in the afternoon. It's going to start to cool off. It's going to crest the mountain and then it's slightly downhill to my place. There's a two-hundred-metre fireguard power line right behind my place. I can stop that fire. That's not a problem.'"

Lee phoned the Cariboo Fire Centre and said, "Come and help me. We can stop that fire. There's no need to let the whole north end of town burn up. They said I was a fool and an idiot to go up there. I said, 'Well no, I'm not. I'm telling you that that fire is quite stoppable and it's just about peaked at cat 6 and it'll be like a cat 2 or 3 in a few hours. Come and help me.' They said, 'No, you are on your own.' I said, 'Okay, I'm on my own and fuck you.' Excuse me, I get crude sometimes, maybe a little too passionate.

"I wasn't putting people in danger," Lee insisted. He phoned a few friends who said they would be there as soon as possible. Pretty soon he had four Cats and then Gord Chipman brought two crews from Alkali. This is why Sam found herself on spark watch on the top of Slater Mountain in the early hours of the morning on July 16.

Sam was not easily put off her game, but she admitted: "I was a little bit scared." She and the rest of the firefighters from Alkali Lake waited until the machines were done knocking trees over, and then they looked for flare-ups and systematically put them out. "These fires were quite small, about the size of a campfire," said Sam. The crew members were able to refill their backpacks and keep going with the help of Lee Todd's old 6 × 6 army truck, a heavy off-road vehicle with three axles and six wheels, carrying a six-thousand-gallon water tank.

While the Alkali crews were putting out sparks, Lee and the other Cat operators built a five-kilometre guard tying into the Soda Creek Road near Blocks R Us, a landscape supply company.

They worked fourteen hours straight. Lee is convinced that if they had arrived even one hour later than they did, they wouldn't have been able to stop the fire. "This whole end of town would have been burnt," he told me. "There's a huge disconnection between the Fire Centre and us as a people. Together no fire could beat us. But there's this thing about them-against-us and us-against-them which is totally unconstructive."

George Abbott and Maureen Chapman wrote a detailed government report about the fires of 2017 that contains recommendations about a host of different issues. In it, they acknowledged the contribution of people like Lee, who worked to limit the spread of wildfires threatening their communities. They proposed that BC develop partnerships with key community members "to provide increased response capacity."[42]

AS THE SKY LIGHTENED, SAM could see something of the lay of the land. Through the smoke she was able to discern a farm with horses, cattle, a donkey—and peacocks. Sam met George Keener, the owner, who had lived there for many years in a small ninety-year-old cabin. He was very grateful for the help he had received, and he told Sam, "You're always welcome here. Come up any time and I'll make you a cup of coffee and something to eat." That the cabin survived was something of a miracle. When Lee had arrived first with his friends, its roof was already on fire and he wasn't at all sure they could save it.

Bridgitte Pinchbeck, the office administrator at Eldorado Logging, laid on a big pancake breakfast for the firefighters at the company office, where a small base had been set up. "Gord came down," Sam told me. "He had a big smile and he's like, 'You guys saved Williams Lake.' I don't know if it's true or not, but the moment when he said that felt really good. Gord doesn't dish out a whole lot of compliments." Sam and the rest of the crew had not slept for twenty-four hours and they were exhausted. But still they were able to muster one more burst of energy. "We were all like, 'High fives,'" Sam said.

Gord told the crew to take some time off and come back at 8:00 in the evening for another session of night ops. When they came back, the donkey treated Sam like an old friend and began following her around. This time, the fires they were putting out were a little larger than the night before, but definitely containable. Sam spent another couple of hours on spark duty, but then Gord called her and said, "Okay, change of plans. You're going back out to Riske Creek. So go home. Go to sleep. Sleep as much as you possibly can and just get out there."

"We went back and slept as long as we could. Then we showed up at Riske Creek and everyone out there started yelling at us. 'Where were you? What have you been doing?' And the other crew bosses and guys that I work with started saying, 'You were supposed to be here at this time,' and I told them, 'Nobody told us that. Actually our boss told us to come here when we had enough sleep. I'm sorry, but we were working all night, you know.' We were like, 'Well, we saved Williams Lake. We were *busy*.'"

BACK IN TOWN, LEE WAS still having arguments with the BC Wildfire Service. Three days after the fire arrived on Slater Mountain, Lee said, an incident commander showed up and told him he was taking over the fire.

"I said, 'Fine. Are you going to have a night shift watching sparks?' The guy blew up. He said, 'I'm a fire behaviour specialist and in my opinion that fire is not likely to cross that guard you built.'"

Then it was Lee's turn to explode: "For the sake of two kids it would cost us eight hundred dollars to run up and down there all night, you'd risk a whole fucking town and five hundred houses?' So I put two of my kids there, actually four of them, two girls and two guys and two quads with water, water on the back, running up and down there all night, just keeping an eye out because the only danger was a spark crossing. They watched every night."

When I talked to Williams Lake mayor Walt Cobb about the log piles north of the city, which never did catch on fire, he said, "We're so fortunate. Two kilometres from the city boundaries, the wind changed or whatever. God was looking after us, I guess. The fire basically just stopped." He wasn't aware that human intervention may also have played a role.

Blake Chipman said that Gord and Lee "didn't wait for any slaps on the back about it and they still to this day don't care about that. They just did what they had to do at that time. Basically what they cared about is saving what they could." It was important to realize that "they were all experienced people," Blake said. "They were not just a bunch of hillbillies grabbing ahold of a garden hose. They knew what they were doing. They got around very close to the edge of the fire and then ran a water truck. They stopped all the hot spots from jumping their fireguard and by morning it was under control."

Sam didn't get to bask in glory during her final stint at Riske Creek either. "It was a lot of infrastructure protection. We were putting out the little smouldering ground fires. And then doing patrol along the highway, sniffing out the smoke and, you know, turning the soil, and making sure it's really out." The house that Sam left on July 15 had survived. "When we went back, we were really surprised that the house was there. It burned all around. It is just so weird the way fire works. You know, one house may burn down and then the neighbour's won't."

Sam and I had been talking for over an hour in the café. We were about to leave but Sam had one more thought she wanted to share. She had actually enjoyed the firefights, she said. "It was exciting and I felt like I was making a difference and helping. Once you get to the end, and you're just on patrol and doing mop-up, then it's like, 'Okay, I'll go back and do my other work.'" (After her second tour on the fires, Sam returned to her regular job.) As an inexperienced firefighter, Sam recalled, "I was always at the bottom of the barrel, but happily it was a good experience."

HANCEVILLE–RISKE CREEK

CHAPTER 13

PHENOMENAL HEART COME OUT

ON A WARM AUGUST MORNING in 2018, Gordon and I drove from Riske Creek along Highway 20 and came to Lee's Corner, where a gas station and historic café once stood. There was not much left. We saw a melted phone booth with a defunct handset hanging off the hook and a burned sign saying "elf erve." We turned left onto the Hanceville Road, drove south, crossed the crystal-clear Chilcotin River, and went down a series of switchbacks. On the right-hand side, we came to the Chilco Ranch, a storied place dating back to the 1880s when an Irishman, Mike Minton, founded it.[43] The ranch, which sprawls over eight thousand deeded hectares,[44] has gone through different owners over the years, many of them "colourful characters." Dean and Lorraine Miller bought the place in 1992; their tenure is one of the longest in the ranch's history.

Today, their adult children and grandchildren have houses on the property. The cluster of residences and barns with their distinctive blue roofs, nestled on a benchland below a ridge, looks like a small hamlet. We had come to hear about the early days of the Hanceville–Riske Creek fire, when the Millers put in an extraordinary effort to save their homes and ranch.

Lorraine, a friendly woman with short-cropped grey hair, showed us into her living room and invited us to sit in one of several reclining chairs. "Sit wherever you like," she said, "but not there." She pointed to one of the chairs. "That's for the chief." When Dean arrived, he sat down in his favourite spot, took off his Stetson and placed it on his knee. A few minutes later, his two grandsons, Justin and Jordan Grier, both dark-haired and bearded, slipped in and sat down. Dean spoke in a quiet husky voice, with much gravitas. Later Gordon said he felt like he was visiting the Cartwright family, of *Bonanza* fame.

This was a four-way interview. Jordan and Justin did most of the talking, but Dean and Lorraine weighed in sometimes to correct or add to what they said. On July 7, lightning hit the south side of the Chilcotin River. It started a blaze, one of a collection of fires that began west of Williams Lake, and which eventually came to be known as the Hanceville–Riske Creek Complex. The fire jumped the river and headed for another ranch the Millers owned on the north side. Justin borrowed a backhoe from some neighbours to try and save their place. He used it to take the fence apart because wooden fences are a conduit for fires. Neighbours showed up to help him, but six and a half hours after the first plume of smoke appeared, the property was burned out. Lee's Corner, farther north, was also gone. "That evening," said Jordan, "we loaded up all of our horses and our tack and took them to Big Creek. We figured we'd be riding around and looking for dead cows. We had two hundred yearlings near Highway 20 that were right in the face of the fire." Dean and Lorraine went to Williams Lake. When they returned on the morning of July 8, the main ranch was still intact.

"The fire sat on the ridge above us, in a perfect line, for a day straight," Justin recalled. But it was raging all around. A red shirt with the BC Wildfire Service, and the only one with a set of cutters, snipped the fence on Highway 20 so the heifers trapped behind it could escape. The next morning, most of them were found near a shuttered sawmill but a few of them had already started moving west along the highway and were coming home. Later the firefighter apologized for wrecking the fence. "But if it weren't for him cutting the fence, I don't think any of them would have survived," Justin said. "It kind of looped up above them, and wrapped them to the highway."

On the morning of July 9, the Hanceville fire was estimated at ten thousand hectares.[45] It had been looming ominously over the Chilco spread for a whole day when suddenly the wind picked up

Dean Miller was well prepared when fire swept down
a ridge onto the Chilco Ranch on July 9.

and the fire swept down from the ridge. "At 11:00 on the ninth, it hit the main ranch," Dean said. Right away, the pumphouse burned to the ground. "Then the power went out and all of our high-pressure water supply was killed," Jordan said. The ranch had an irrigation ditch, but for a few days it, too, was out of commission. The river, of course, was a source of water but it was a couple of kilometres away, not an ideal situation when dealing with a fire as powerful as this one.

Jordan said, "Once it came here, there were some forestry guys trying to help me fight the fire on the hill with a few pieces of machinery. Then those guys left and went to the hayfield. There were quite a few people here, maybe about eight, fighting and spraying water on everything. I went to my house, the next one along here, and it was already on fire right next to the driveway and we didn't see it, there was too much smoke. So I headed up the driveway and a couple of neighbours showed up with big water tanks and we pulled them up to the top of the hill and we started fighting fire down into where my house is. Machinery was on fire and tires blowing up everywhere." Justin added, "I ran over to the forestry trucks and grabbed every bit of hose they had to keep us going here. One guy jumped in with me and helped."

"Everybody in the field there thought we were all dead," Jordan said. "They thought we were going to have 'er. We were ripping my shop apart because it was on fire inside the walls. With a pair of bolt cutters, we were pulling the boards off." While the two brothers were desperately trying to save the shop, they looked up toward Justin's place for a minute, but there was so much smoke they couldn't see his house at all. Justin recalled, "I told Jordan, 'I think my place is gone.' We chuckled and I said, 'We might as well try to save yours.' 'Cause you can't do anything at that point."

And then the smoke lifted briefly and revealed that Justin's house was still standing, although mightily beleaguered. Four BC Wildfire helicopters showed up at exactly the same moment that fire hit the house. Buzzing back and forth like mad bees between the river and the ranch, they scooped and dropped, scooped and dropped. Justin timed their trips, about a minute each way. It was touch and go; the lids on his septic tank were burned. Sitting in the Millers' living room Justin gestured to an air purifier sitting a couple of metres away. "That's how close it was," he said. "If it weren't for those helicopters, my house would not be standing."

The fire's ravenous appetite was not yet appeased. Janet Miller, Jordan and Justin's mom, had a stunning log home on the property as well. (You can see pictures of it in the May 12, 2015, issue of the *Western Horse Review*.[46]) When a small, detached guest cottage nearby went up, the fire's wind blew the debris onto the side of Janet's house. Luckily a couple of quick-thinking firefighters saw what was happening, jumped out of their trucks and dragged some fire hoses through the house and onto the deck, from which they were able to extinguish the insidious embers. Jordan credits the two men with saving the house; without them the log structure would surely have succumbed.

Even after the blaze rolled through the main ranch houses and

Justin and Jordan Grier, Dean Miller's grandsons, were up for four days straight defending their ranch.

buildings, Jordan and Justin did not rest. Hot spots kept popping up here and there as a gust of wind would catch a slumbering ember and rouse it to open flame. Trying to prevent the conflagration from starting up again, they dashed from one small fire to the next. Fortunately, they had help with the patrol. At one point, when Jordan checked on his own house he found a "local guy" there, standing on the front lawn with the garden hose running. The fire was out and he gave Jordan a thumbs-up. Satisfied that the crew at his place were in control, Jordan said, "Thank you," as he flew over to another problem.

At the beginning of the fire, Jordan recalled, "I think we were up for four days straight and no sleep."

"The best feeling would probably be about the ninth," Justin said. "We were right in front of the barn where we had the low-bed [a type of trailer] parked. It was 3:00 in the morning and I just lay on the neck of the low-bed with my feet stretched out and I finally got to take my boots off. I took a picture because that was the best thing in three days." He also mentioned enjoying a cold, burned smoky that he found lying on a table in his grandparents' house. "I finally got to eat one!" he said.

"Yeah," Jordan said, also remembering what it was like. "They'd be cooking and you'd come in to get something and the second you sat down, you'd have to leave. There was something burning and you had to go."

This was a time for circling the wagons. For security, the family all moved into the "big house"—Dean and Lorraine's place. "So we knew where everybody was," Lorraine explained. "We had cats and dogs and all the family. If somebody didn't show, and we didn't know where everybody was, we'd go looking." They did this for at least a week. "You didn't know," Jordan said, "whether your house would burn down. You'd go back and even with it being wet, it would smoke right up. There'd be sparks going everywhere as soon as you got wind. At least here, we knew it was safe, everybody was here."

Lorraine set up a camp kitchen just outside the back door. In the morning, she'd serve breakfast sandwiches. "We called them Miller Egg McMuffins and anybody who would be coming over, we'd give them a muffin and a cup of coffee," Lorraine recalled. The RCMP stopped in too and grabbed some breakfast before hitting the checkpoints.

The Chilco Ranch became the command centre for the local firefight, as well as something of a safe haven. The family stowed its equipment on an irrigated field, the safest spot for it. They also let their neighbours bring in their machinery: trailers, boats, ATVs, tankers, tractors, cats, low-beds. Justin welcomed one and all. But he had a rule: "'The key better be in it, it better be full, and you're showing me how to run it if you're going home.' As simple as that. We got some tutorials on a couple of pieces." One night, a hay bale caught fire, and the old Cat belonging to the Chilco Ranch didn't have any lights, so Justin borrowed one of the neighbours' Cats parked on the field. He quickly figured out how to use it, rushed over to the hay bale, pushed it around until the fire was out and brought the Cat back.

Certainly there were losses. Some were easier to recover from than others. A lot of hay caught fire and some of the stockyards burned. The ranch lost its domestic water for the houses and irrigation for the fields. Jordan resurrected both, no easy task. The machine shed was destroyed, and machinery as well. Fences needed to be replaced and rangeland was damaged. The Millers' other ranch, which burned on the north side of the Chilcotin River, is now vulnerable to flooding. Scorched ground is unable to absorb falling rain. "Now that the fire's been through there, every time it rains, it floods and it's my problem," Jordan said. "Everybody wants to pass the buck. You get a torrential rain like we had over there recently and the whole hayfield is a writeoff. There are logs in the middle and everything washes down on it. You rebuild it, but if it happens again, then what? The ground is hydrophobic until something grows on it." While the provincial government has programs to support the rebuilding of fences, so far Jordan has not been able to find anything to help him with restoration of land flooded as a result of the wildfire.

On the positive side of the ledger, Cat operators were able to save the barn with a savvy use of firequards. The main buildings, other than the small guest cabin, were rescued. Most of the livestock survived. "I can't believe how much we saved," Jordan said. "Once the wall of the fire is through, it's just a matter of keeping everything wet for as long as you can." A team of firefighters came to help with that. "I will say, the cleanup crew, who were putting out hot spots around the place, did a fantastic job," Dean said. In the months that followed, the rangeland started to come back, although Dean pointed out, "It takes a few years to establish a solid grass mat." The government replaced forty kilometres of fence along the highway for the Chilco Ranch to keep the cows contained, off the road and safe.

Fire is no stranger to people in the Chilcotin. Dean and his family had been anticipating it for some time. He said, "You know you live in a fire area. Let me rephrase that: you know you live in timber country, you better be prepared for fire. We were in a real dry belt here so we had our own firefighting equipment. There's nobody better than yourself to be a first responder, if you're capable and able to do it. We are. With a Cat, excavator and low-beds around, we can move equipment. We've got a 5,500-gallon water tanker, which we were fortunate to buy when we did. We use it strictly for firefighting."

"And dump trucks, tractors," Justin interjected.

"I had nine or ten pumps," Dean added, "thousands of feet of hose, and nozzles. I went and bought more in early March."

Lorraine said, "All of our tanks are filled with water, in case."

After the summer of 2017, Dean bought a couple of containers because his machine shed, where he normally stores things, burned. One of the new containers is now set up to store firefighting equipment exclusively, with hoses, pumps and generators. Preparedness is a way of life. But it is also more than that. It is a deeply felt ethos. "You get used to being an hour from town," Jordan said. "It's not a big deal for us. You get the wrong type of people living where we do, they wouldn't have a chance. We're all very capable. We weren't going to run out of food; we had enough stored for another three months." Jordan knows how to wire a generator to a house so it can run on full power. When the hydro went out, it was not a problem; he could keep fridges and freezers going. He maintained, "If you live in an apartment and the water is turned off, you phone the guy. That's how most of the country is these days. The survival instinct of people isn't there anymore. People don't know where their food comes from. They're out of touch."

"They'd Google it, ask Siri," Justin added with a grin.

"Most of it comes down to common sense," Jordan said. "You live at the end of a dead-end road, nobody can help you. You've got to be set up. The people who can be doing this stuff, like us and every other rancher that has the technology and the know-how, should be doing it." Jordan was reminded of the guys who were helping him fight the fire on the hill on the first day it hit the ranch. "They were crying," he said. "They were in tears, totally scared. Most of the people that live out here are not scared to get in there. But if you keep getting this drilled into your head all your life, that you can't do this stuff, people forget what they can do. When there's nobody around to tell you that you can't, you would be surprised what you can do."

And then Justin told a story that illustrated Jordan's point perfectly: "I went up to one local guy in a water truck. A fitting broke on his pump so he's gravity watering. I said, 'What do you need?' He told me and I said, 'Oh we just used it last night.' So I came down, started a generator, started a welder, cut everything apart, welded it up, went up, put it on. He couldn't believe that I made him a fitting for his pump. Three hours later, he was the very first one in the middle of the flames, saving our stuff. Phenomenal heart come out."

Jordan said, "We're very fortunate and outfitted so we kept a lot of our own stuff going and a lot going for other people. As soon as we were okay here and the sprinklers were on after the fire went through, we weren't home. We were up and down the highway with the neighbours and fire tanks. We didn't register with Forestry, although they asked us to. But I told them we wouldn't. If you register with Forestry, then they pay you to go and fight fires. But then they also dictate what you can and cannot do."

"It's called going rogue," Justin said.

"Whatever you want to call it," Jordan said, "most of the guys that we know out here have no problem with it." The fires left an indelible mark and a strong feeling of community. Jordan said, "I never say no to anybody. If somebody stops to talk to you, you make the time. It's important. It's amazing how important that is."

"There's a positive in it," Dean added. "Everybody stepped up to the plate. You can criticize this person or that person, but really the positivity of the people who stepped up to the plate helped to defend us."

It was bracing to be with people who were so independent. Such attitudes are not valued or cultivated so much in the city. But with climate change coming, and more disasters, floods, storms and fires forecast, I wondered whether we should do more to encourage them.

As Gordon and I drove east once again, we came upon a family driving cattle. Two boys, about the age of ten or twelve, a woman and a couple of men were all on horseback, their saddles fitted out with lassos. They had four dogs with them as well. I was surprised by how quickly the group moved, the cattle at the end of the herd panting, puffing and struggling to keep up. I also remember how the riders shouted to the dogs. One in particular, Trigger, perhaps new to the game, seemed to need a lot of direction to do the right thing. The other three kept easily to their appointed places and tasks. I enjoyed seeing this co-operation between the people and the animals. We followed along for a couple of kilometres and then the drive moved off into the trees on the left-hand side of the road, but I felt privileged to have seen something so graceful.

CHAPTER 14

THE FIRE DRAWS EVERYTHING TOWARD IT

"YOU BETTER GO," MIKE JASPER said. It was one in the morning, but Mike thought he'd better stop by his sister's place and warn her anyway. He had been driving around Riske Creek all night looking for flashpoints when he saw fire jump Highway 20 and reach the garbage dump. The flames were uncomfortably close.

I was sitting at Noreen McDonald's kitchen table beside a massive stone fireplace, drinking coffee and listening while she and her husband, Ed, talked about the summer of 2017. Noreen told me that fire had first arrived in their area on July 7 but burned fairly far west of them. Her brother had knocked on her door when the winds flared up a second time. That night, she and Ed spent a couple of hours packing, then drove to Williams Lake, where they stayed with their son, Jason.

Almost as soon as they were out of their home, they began worrying about it and the fact that no one was there to protect it. Three days after their departure, they returned. It was lucky, Noreen said, that they were in a tow truck. The police let them through the checkpoint at the Fraser River Bridge because Ed was the owner of Chilcotin Towing, based where they lived. He was the only person around capable of providing emergency towing. He was an essential service.

Ed has lived in Riske Creek since 1975, when he came to visit a friend and never left. Noreen has been there for all but two of her sixty-five years. Those she spent not far away, in Williams Lake. Her grandfather, Wes Jasper, came to the area from Kentucky in the early 1920s. Her father, Delmer Jasper, was born in 1925 at Meldrum Creek, just forty kilometres away. Both Wes and Delmer loved horses and established the Milk Ranch; they had cows and raised sheep. Noreen and her seven siblings grew up in the house that her grandfather built on the ranch, living the way pioneers did. "When we were kids, we used to heat our water to bathe and we all used the same bathwater. There were four or five kids in one bedroom. We didn't have hydro or TV. Now times are totally different," Noreen said. But the family is still a mainstay of the community. Three of Noreen's brothers, Mike, Pat and Leland, live there. Mike and Pat work as cowboys and were inducted into the BC Cowboy Hall of Fame in April 2018.

On July 14, there was another knock on Noreen's door—an RCMP officer. "The wind is going to be bad tomorrow. It's coming," he predicted. This time, Noreen had no intention of going anywhere. "For me it was like, 'Let's get it over with.'" The next day, around 11:00 in the morning, Noreen's sister-in-law, Lorraine, called. Her place is about ten kilometres away, where the Milk Ranch and Noreen's childhood home used to be. She was alone and the fire was on its way, she said. She wondered if Noreen could help her move the dogs—all nine of them.

"Yes, I'll be right there," Noreen assured her. She was alone too. Ed would have come with her, but he was in Williams Lake with their son.

Noreen recalled, "I go up the road about a mile to a place we call the Big Rock. The cops were there and had the road blocked off. There's a house there and two helicopters were dropping water. They wouldn't let us through." Noreen talked to one of the officers and explained that she wanted to help her sister-in-law farther west along the highway. "Don't worry about it, I'll go back and give her a hand," the officer promised. Noreen turned around and headed home, but she didn't make it. When she got close to the garbage dump, about a kilometre and a half from her place, she saw a blaze on the property of some neighbours, Ken and Tracey. "I was scared," Noreen told me. She ran into their place, yelling, "Oh my God, it's here."

The scene was chaotic. While Ken was falling trees to clear a break, Tracey was fighting the fire with a hose attached to a water tank on a truck. Noreen grabbed a couple of two-gallon pink and green buckets from her vehicle and used them to douse flames. "What do you do?" she asked. "You gotta help." She laughed at the memory of herself tackling a wildfire with those incongruous pink and green buckets. Two other ladies showed up to help and Noreen began moving antiques out of the house. Then the fickle wind veered and pushed the fire toward a creek. More neighbours arrived, and finally, a crew of firefighters.

While this spirited defence was underway, cars were barrelling along the highway at a great speed, even though it was smoky and the visibility was poor. One of the firefighters asked Noreen if she could flag the vehicles to make them slow down. She directed traffic for a while and then noticed flames leaping south to her side of the highway. After observing them for a few minutes, she handed the "Go Slow" sign to one of the firefighters "That's my house over there," she said, pointing. "I've got to get home."

When I asked Noreen if she was ever concerned about getting burned, she said, "It's funny. I was really scared at Ken and Tracey's. It was going through my mind. 'Ed's not here. Jason's not here. There are no firefighters. We don't have any bladders set up.' It was just like, 'Okay, I'm on my own with the dog.' But when I told that guy 'I'm going home,' I was calm." Lynn Bonner, a friend of hers, was surprised by how unperturbed she seemed to be. 'How can you be like that?' Lynn asked. Noreen was a bit surprised by herself at the time. She thought maybe the panic would kick in later.

Just as Noreen got to her place, Ed, Jason and his girlfriend, Alena Mayer, arrived. Hytest Timber, a local logging company, appeared with low-beds and Cats to put fireguards around the McDonalds' property. As well as the house, the McDonalds had a barn, a shop, a pumphouse, a fuel tank, a couple of chicken coops and some sheds. A yard filled with used cars was in front near the highway. Ed, who has owned his towing business for twenty

Ed and Noreen McDonald, owners of Chilcotin Towing, saw their collection of thirty-three cars succumb to a rank 6 fire.

years, acquired the vehicles in various ways. Some were damaged in traffic accidents; others were impounded by the police because they were stolen, or abandoned and unclaimed. Ed kept them for their salvage value.

A crew of firefighters from Terrace set up bladders. Ed and Jason helped them to install sprinklers on the roof. "We tied one to a chimney and put another at the other end," Ed recalled. "We know that when the fire is going, it draws everything toward it. You don't want to put a sprinkler just on one end because it would be sucking the water away from the house." By the time the sprinklers were in place, the fire was practically on the McDonalds' doorstep.

Noreen had nothing but praise for the firefighters from Terrace. "They were really, really good guys, really, really on the ball," she said. Blake Chipman also liked them. He told me they were not only on the ball, but ballsy. Luckily, Noreen's worry that she would be battling the fire with only her dog as a companion did not materialize.

Noreen and Alena had been spraying the lawn and the house, but the firefighters took over that task. "You've done what you can, it's time for you to leave," they said. Noreen grabbed her dog, but the cat was nowhere to be seen. It always slept in one of used vehicles and now it was too late to look for it. The firefighters started yelling at Alena and Noreen: "Get out of here!" As they ran down the driveway, the firefighters shouted, "Watch the hydro lines. Don't fall."

The sight of the fire licking its way through the property had attracted a crowd. About fifty people were standing on the north side of road watching it. Human beings have had a relationship with fire for a million years or more. During that long association, we've learned to tame it. Every day we light fires. We have furnaces in our homes. Our vehicles are propelled by tiny explosions that go off hundreds of times per minute. We even take to the sky, airborne on jets of burning gases. But rarely do we see what the residents of Riske Creek witnessed that day—the raw force of a fire, undirected and unharnessed.

Experts talk about the "frontal energy" of a wildfire, the energy expended over one metre at the fire's front. When a fire is roaring through the crowns of trees and flames are shooting up for fifty metres or so, the intensity can hit 150,000 kilowatts per metre.[47] In one hour, a metre of fire releases enough energy to power fifteen homes for a year. The throng of people at the side of the highway were mesmerized by a majestic power.[48]

The McDonalds joined the onlookers and saw Ed's collection of used cars succumb to the inferno. Ed remembered, "When those vehicles started to go, it was like World War III. The gas tanks, the tires would go, even the drive shafts, because they are a sealed unit. The roar was unreal, just like a big blowtorch." One of Ed's neighbours said, "I think your house is on fire." When the thick smoke cleared a bit, Ed could see that it wasn't the house, but he didn't know what it was until afterwards when he discovered that a truck behind his place had burned.

The firefighters let Jason stay behind and help, so he sprayed down the shop. "We didn't even know he was there until the smoke opened a little and we could see him over there," Noreen said. And then the smoke closed in again. Noreen no longer felt so steady. "I was just about to cry," she told me. "I said to a firefighter, 'Our kid is over there. Where is he?' He was, 'He's okay, he's okay.'" When the smoke cleared again, Noreen realized that Jason was now spraying the tractor. He was drenching the machinery they had moved to the middle of the field.

The fire was selective about what it consumed. While all thirty-three vehicles in the front yard burned, the fuel tank was, mercifully, untouched. And a circle of cars behind the home was disdained except for the one white pickup in the middle. "The grass that burnt went like a snake around all those other vehicles and got to it," Ed said. This was what his neighbour had seen burning. "A two-year-old truck with not a mark on it—except for the tires," Ed recalled, shaking his head ruefully.

Thanks to Jason's efforts, the shop survived, although there was no paint left on one side of the building. He'd rescued the tractor, too. But the pumphouse went, along with the willows in the creek. Ed said, "What hurt was that we lost a bunch of stuff in the sheds from my dad. He had lots of old lanterns, coal oil lanterns." The kitty was never found.

The fire burned around the house on both sides. It was so close that it scorched the telephone lines and the McDonalds couldn't make any calls for eight days. But the house itself was not singed. In all, it didn't take long for the fire to go through. Not even an hour passed before it was over. It tore over the hill and when the folks watching the display couldn't see it anymore, they dispersed.

Noreen told me that when she was a kid, "We'd have a fire every summer. Dad would go to burn the grass and burn Mom's greenhouse down. He'd have to rebuild it—*every year*. She got so mad," Noreen recalled, laughing. "'Don't you burn the grass this year,' she'd say. He'd shrug and say, 'It'll be okay.'" The McDonalds' shop had burned down one year, due to a propane explosion, but the fire didn't spread. These were manageable fires, not like the ones of the summer of 2017. They were something new, Ed said, not like the old days at all. These fires blew hotter, spread faster and went farther than anything anyone had experienced before.

When I asked Noreen what she felt like when she discovered her house had been spared, she said, "I can't explain it. It was just weird. When the firefighters told us we could go back, we are walking up the driveway, and we see our house is here. The log barn and the chicken houses are gone, but we're here, Jason is here and the house is here. 'Carry on,' I thought. 'You just have to keep going.'"

The firefighters left the McDonalds a bladder, a pump and hoses. Even six to eight months later, Ed could see wisps of smoke spiralling up from the ground. Finally he went in with a machine and dug up some burning roots, a hidden menace. The McDonalds got nothing for their cars, although their contents insurance covered a portion of their other losses. Ed and Noreen are planning to rebuild some of the structures after the fences are replaced and are waiting for fallers to carry away the charred trees. "They're logging a lot of burnt wood around here," Ed said. "They have a few years to get it. Otherwise it dries out too much and there's too much crack in it."

The McDonalds found that cleanup takes a long time and there were after-effects they didn't think about; for instance, the ruin of their buildings left the land littered with nails. They had to buy a magnet to pick them up so they wouldn't get flat tires.

The spectacle of the burned cars was another kind of magnet. For a long time, Ed said, people were snapping shots of it. "They would drive in, take pictures and drive out again," Noreen said. "Sometimes a car on the highway would stop, the window would roll down, you'd see a camera with a very long lens. They knew what they were doing."

Jesaja Class, a young photographer (and magician) from the Nemiah Valley was one of the people who took a picture of this icon of destruction. In it, you can see the pitted, heat-warped metal, twisted and distorted shapes, shattered glass, rust and sepia-obliterated colours. Cars have been a potent symbol of our age with their promise of freedom and power. Behind the wheel, you feel like the master of your fate. No wonder there are a billion cars on the world's roads today. To see them utterly wasted is sobering and compelling—like looking at the graveyard of a dream.

CHAPTER 15

BUT IT CAME, IT CAME, IT CAME, IT CAME, IT CAME

KURT VAN EMBER IS A tall man with an affable hostly manner. In April 2017, when he and his wife Brenda purchased the Chilcotin Lodge in Riske Creek, it seemed a remarkably good fit for them. They both love history; Kurt collects antiques and has restored vintage cars as a hobby. The lodge, a distinctive green-and-brown shake-and-log building set on a wide, grassy slope, was built in 1940 and is still furnished with period pieces from that era. When Gordon and I arrived on a cold sunny day in February 2018, Kurt went down to the basement to fire up the furnace—heated with wood, as it was in the old days. In 1944, the Canadian navy commandeered the lodge for a wireless relay station, probably to intercept Japanese radio communications. In the lounge, Kurt proudly showed me a radio that was the same model as the one the navy had set up during the war.

But there must have been moments during the summer of 2017 when Kurt wondered whatever possessed him to take on such a venture. As he recalled, "Riske Creek was the centre of the universe for the fire. It came from the south, it came from the west, it came from the north. The only place it didn't come from was the east. You look at the dos and don'ts of running a new business. This whole fire thing was never in any document."

On July 7, as Kurt and Brenda came home from the Williams Lake Stampede, they saw the lightning strikes to the west. Kurt said to Brenda, "This is not going to be good. It is so dry here. We are going to be in for one hell of a ride." (They'd had nine millimetres of rain since the beginning of June, a fraction of the sixty-eight millimetres that is the historical average.)

Brenda left for Ontario to be with their daughter, who was about to give birth, but her eighty-year-old dad was still at the lodge visiting. "There was smoke everywhere," Kurt recalled, "and we didn't really know what was going on." He urged his father-in-law to be safe and go home.

Brenda and Kurt Van Ember, proprietors of the Chilcotin Lodge, were new to Riske Creek when fires threatened their business and home.

Kurt's twenty-three-year-old son, Chris, a cheerful young man, was at the lodge too, but he had been working there and his plan was to stay and help his father. On July 8, he went into Williams Lake to get supplies. Anxiety was in the air. The Walmart parking lot was jammed with RVs. Many people had the same idea—to stock up while they could. All the plain bottled water in the store was gone but Chris could buy carbonated. He drove back to Riske Creek without problems. The next day, on July 9, an evacuation order was issued and the roads were closed—with reason, he said as we chatted.

Chris showed me the pictures he took when the evacuation order was issued. I could see the crimson clouds and the plumes of smoke to the west getting larger and larger as he clicked through the images on his phone. But then the fire was still twelve kilometres or so away; Kurt and Chris were not in any immediate danger. They spent the day rebuilding a deck at the front of the building. Taking the precaution of gathering up the wood scraps and chips generated by their work, they scattered them in a far field. They might still ignite, but at least there they didn't pose such a threat to the lodge.

That evening when they were finished, Kurt felt celebratory. He said to Chris, "We're going to have a steak, a glass of wine on our new deck. If we only get one steak, one glass of wine, that's all it's going to be. But we're going to do it." The red glow in the west was getting bigger and bigger. "The first time we saw flame," Kurt said, "we were eating our steak. It was unreal."

Highway 20 goes past the lodge and then angles northwest. When the fire broke over a hill a couple of kilometres to the west, it was on the south side of the road but still north of the lodge. Kurt noted the direction of the wind—from the northwest—and thought, *We're done.* But he decided to have a closer look. He didn't have much in the way of firefighting equipment, but he had two hand-pressurized water tanks, so he grabbed those, and he and Chris rushed over to the blaze in their vehicles. "The smoke was rich and thick," Kurt recalled. "We got to the top of the hill and everybody was there. We started beating it down. All the neighbours came. I didn't know half these people. I didn't know who they were. Everybody had water trucks and skidders. We worked our butts off to keep that fire from jumping to the north side of the highway, because if it jumped to the north side, it would circle around and eat everybody. We managed to do that. There must have been fifteen, twenty people, from nowhere." The fire abated about 11:00 p.m., and Chris and Kurt quit for the night.

On July 10, Brian Fuller, a neighbour, came over and dropped off a pump, a fire hose and an eight-hundred-gallon water tank that he knew weren't being used. "I couldn't believe it," Kurt said. "That was a blessing." Tanks were suddenly in short supply. "You couldn't pick up a phone and order a tank, so that was a real sign of life." Kurt thought, *Wow, maybe I might have a fighting chance here.*

Brian was an ex-logger, a heavy equipment operator and the site manager at the Old School, the former elementary school in Riske Creek. (The Toosey First Nation had converted it into an education centre for adults that offered training in building houses, operating sawmill equipment, being a care aide, and more. In July 2017, it got a contract with the Cariboo Fire Centre to open a fire camp on the grounds and used the profits to expand its curriculum.)

The Van Embers had gotten to know Brian in the three months they'd been living in Riske Creek and he had become a friend. At Kurt's suggestion, I went to visit Brian at the school, about a kilometre up Stack Valley Road from the Chilcotin Lodge. A bear of a man, Brian was well versed in the ways of fires. He told me that you had to be strategic. "Once that fire drops down from the crown and gets to the ground, then you can manage it a lot easier. You can put it out with fireguards. But if it's a full-blown hundred feet high, just get out of the way. It's too hard, too dangerous." He was wise in the psychology of firefighting too. He said, "Kurt is a good example of somebody that was really having nightmares. I said to him, 'Drive up the road, you'll see the fires right there.' Everybody has this image of a fire, that it's so crazy—dramatized sometimes on the media. I told him, 'Go see it when the wind has died down, especially in these areas where there's a lot of grassland, where you can manage it pretty quick with dozers. Just drive up there for five minutes, go take a look and see what the fire is and come back.' It wasn't as crazy as he imagined."

Brian had already used a dozer to put a guard around the school, but he was worried about the fire farther west, where efforts to contain it were in danger of being overwhelmed. The day

Brian Fuller, an ex-logger and heavy equipment operator, helped to keep the Old School in Riske Creek safe.

he dropped off the pump and other supplies, he also came to get Chris. "I need you, so let's go."

"Okay," Chris said. He had no training in fire suppression, but he was young and energetic. Brian needed bodies," he explained to me, shrugging. When Chris hopped into Brian's truck, it was 8:00 or 9:00 in the morning. Chris was wearing a t-shirt and sweatpants, not exactly regulation gear for the task ahead of him, but Brian paid no heed. He drove forty kilometres west on Highway 20 and dropped Chris off at a spot behind Lee's Corner.

"When I got there they had just finished the awesome eight lanes of fireguard that was to protect Riske Creek," Chris said. "The guard was in an open field so flat," he remembered, "you could literally land your plane on it." The idea was to stop the fire and prevent the grass from burning and setting the trees on fire. "But it still jumped," Chris said. "When I got there, they were trying to put out the fire in the gully. I was with some guys who knew what they were doing. They had an eight-hundred-gallon water tank and were sucking water from the gully and spraying with the hoses. They gave me a high-vis vest and a piss tank. I would follow them around and put out spot fires."

Meanwhile, back at the lodge, "The phone was ringing off the wall," Kurt said. "Everybody thought, *doomsday*, right? It was just a matter of chaos, pandemonium." Brenda's dad was phoning every day, every half-day. Brenda was phoning every twenty minutes. Kurt also spoke to his dad, a vigorous seventy-nine-year-old who still flew his own plane and who lived in Chilliwack. He told him, "Kurt, you're in a hell of a spot. I'm not going to tell you what to do. You do what you need to do. I've been in your situation before in Alberta. The fire came on me and I was trapped. And the only thing I could think of was to dig a hole with a flap of muskeg. I got into that and pulled the muskeg over me. The fire rolled over me."

"Well, Dad, there's not much muskeg up here."

"Yeah, you're kind of hooped."

"Yeah, I know."

"I'll pick you up."

"Dad, it's a no-fly zone."

"I'll come and get you, if you need me."

Then two RCMP officers came to the door. "Who's the owner?" one asked.

Kurt said, "You're looking at him."

"What's your name? How many people are on-site?"

"My son and I are here. He's not here at the moment. He's fighting fires."

"We have an evacuation order. You have to leave."

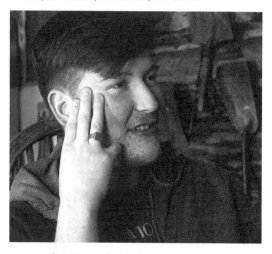

Chris Van Ember had no training in fire suppression when he was "drafted," handed a "piss can" and told to extinguish blazes.

Kurt had never been under evacuation order. But he wanted to stay to protect his place. And he was convinced that if necessary he could leave quickly. He had already packed his prints, china, sculptures he'd made, and family photos. His vehicles were pointed toward the road and had keys in them in case they had to go. His dad, who was on standby in Chilliwack, had access to five airplanes and could come and get them out if the roads were blocked.

Kurt is quite tall himself, but the officer was even taller. Kurt looked up at him and said, "Well, there's two of you and one of me. I'll lose, but I'm not leaving without a fight. I'm not leaving without my son. He's out fighting fires and I don't want him coming back and finding I'm not here. I ain't leaving."

"Calm down, relax. Okay, we're not here to physically remove you. We can't remove you from your building, your property, unless you're a child. How old is your son?" one of the officers asked.

"Twenty-three."

"I can't remove your son. But if you leave your property, I will remove you."

"If I leave my property because I'm fighting fires, how do you feel about that?"

"We'll look the other way. If you're just cruising around, I'll get you."

"Fair enough," Kurt said.

And so Kurt was left to fend for himself. When Chris came home, his father took a picture. Chris said, "I did look like hell. I was wearing a white shirt which kind of turned to black and my face was pretty smoky." Brian didn't pick him up again after that. Other trained firefighters arrived and for a few days, the winds died down, making the situation somewhat less critical.

But on July 15, the fire roared back to life. In the evening, things looked grim to Kurt and he decided to release Indy, a horse they were caring for. Chris didn't think the horse, with his black coat, would make it. The visibility was terrible at the time, the smoke impenetrable. Everyone was in a hurry and might easily miss seeing him on the road. Chris took the safety vest he had been wearing firefighting and tied it around Indy's neck. He opened the gate, let Indy go and hoped for the best. An hour later, Indy trotted back up the road. The Van Embers decided Indy didn't want to be on his own. "He's a great horse," Kurt said affectionately. Their next-door neighbour, David, who had a horse trailer, offered to look after him.

Close to midnight, the fire popped over a nearby hill. "You could feel the heat instantly." Chris said, "It suddenly went up ten degrees." The morning of July 16, the fire crossed the highway and started heading toward the lodge. "I knew that if the fire and the wind were bad, I wasn't going to be able to save the building. I knew it, you just can't. It's too much," Kurt recalled. But again the neighbours rallied. "We're all beating it down," Kurt said. This time he had the water tank Brian had given him, which he made mobile by loading it onto his tractor. "We're going down the hill, beating this thing out. It was really, really stressful. It jumped the highway, right there." He pointed out the window to a spot in the field in front of the lodge, about twelve metres away.

On this round, Chris said, "Helicopters were flying all over doing their thing." There wasn't much communication between those in the air and on the ground. But when Indy's hay bale caught fire, around 2:00 in the afternoon, some firefighters who were walking along the road saw what was happening and called a BC Wildfire

Service helicopter with their radio. "They had to drop two water buckets from the helicopter onto the hay bale, so we could get it out." Chris said. "The burning hay bale had the weirdest sweet smell. I don't think I'll ever forget it."

In general, Kurt observed, "We were doing very poorly. We're all just yelling and screaming. You're busy doing the fire right in front of you, and then it's beside you and right over there. You're doing that. This guy is fifteen feet away doing the same thing. So everybody is really focused on what they're doing. And all I heard was 'behind you'; everybody is yelling 'behind you.' Everybody was doing that because it was getting out of hand. So it was coming up the hill. How do you fight this? There were pieces of trees and branches and debris the size of the lampshades coming down. The fire creates its own storm. People think it's just grass and that you're going to be okay. You're not going to be okay because there's too much of everything. Too much dry grass. Too much wind."

Still Kurt thought, "Well, I'm not leaving this building that's been here since 1940—not without a fight. It belongs here. Too much history here to say, 'I'm outa here.' I couldn't do it. Can you imagine if I would have left this building and all these local people fighting? I would never have come back. I couldn't ... Instantly everybody was like this." And Kurt crossed his fingers to show just how tight the community had become. "The Chilcotin people stuck together and they fought. It was just amazing. You want to get to know your neighbours ...

"I think the only thing that saved the lodge that day was the wind," Kurt said. "There was a moment where the wind changed. You're just so busy. You don't see anything different. Then all of a sudden the wind backtracked on itself and stopped coming toward us. The fire went back on the grass that was already burned and kind of fizzled out. If it wasn't for the wind change, I know for a fact that we wouldn't be sitting here."

Chris and Kurt called it a night around 11:00. They were covered in soot and worn out. Chris fell into bed and Kurt went to have a shower. When he came out, he saw blue and red lights flashing outside. He thought, "Here we go again. No sleep." Two fire engines and three smaller fire trucks were coming down the laneway. He walked outside to see what was going on.

"You have half an hour," said one of the firemen.

"Half an hour?" Kurt asked.

"The fire is coming from the north now." Kurt looked and sure enough, the whole north was aglow. He'd been so busy fighting the fire coming up the hill to his place from the south that he hadn't seen it.

"I'm telling you," the fireman said. "We will not be able to save this place."

"Why are you here?" Kurt asked.

"We are going to try and save all those houses next to the tree line. If we can stop it there, we can stop it here. If we can't stop it there ... "

That night the fire never came over the hill. The wind changed direction again and blew out of the south. At 7:30 in the morning, after being up for about thirty hours, Kurt served the firefighters cinnamon buns and coffee. They were dispatched back to Williams Lake, but assured Kurt that a sprinkler crew was coming.

On July 18, as promised, that team, a local First Nations group, arrived. Its members did not protect dwellings, but the area around them. They were specialists in ground firefighting. They started poking sprinklers into the earth, connecting all the lines together and attaching them to Kurt's pump. While they were doing this, a huge truck with a trailer came into the yard: more firefighters were coming on deck.

Kurt asked one of the new arrivals, "What's up?"

"We're the sprinkler crew."

Kurt pointed to the folks who were already working at his place. "We have a sprinkler crew here."

"We are the *structural* crew," the firefighter explained. "Those guys are the *ground* crew." He explained that the members of his group would be putting sprinklers on the roof of the lodge and attaching them to an eight-hundred-gallon bladder they had brought with them and that they would be filling with water.

"What about my neighbour?" Kurt asked, gesturing to the house on the east side of his place. He knew that David, whose father had bought the property next door, had lived there a very long time. The structural specialist said he would check it out. When he came back, he told Kurt, "We're doing David's place too."

Now Kurt had ground and structural protection. He relaxed a little and all the attention he was unexpectedly receiving made him feel a bit giddy, so he decided to have some fun with the bladder. On a piece of paper, he wrote, "Chilcotin Lodge Management. No Diving." Then he drew a picture of a diver in a circle covered with a X. And then, "Pool Closes at 10. Management Staff." He held the paper against the side of the bladder and taped it on. "People got a kick out of it," Kurt said. "Somebody put it on Facebook."

On August 4, Chris felt like taking a break and went to see his girlfriend in Salmon Arm. Kurt's wife, Brenda, came back from Ontario and the firefight dragged on. The lodge lost power a few times, and in Riske Creek a big effort was invested in protecting a communications tower critical for landlines, cell phones, the internet and data services in a large area stretching from Riske Creek to Bella Coola. The fire came close to the lodge property one more time—this time from the neighbour's field. "We put that out," Kurt said. And then on August 12, high winds started stoking the fire in several places and an evacuation order was issued—again—for properties all along Highway 20 from Alexis Creek to Riske Creek. But by August 15, cooler conditions had returned and the order was rescinded. Chris came back to Riske Creek and worked as a chef for the fire camp in the Old School and the larger BC Wildfire camp 1.8 kilometres away from the lodge.

A semblance of normalcy gradually returned, but the fire kicked up one last time. Chris remembered that on the evening of either September 11 or 12, he was sitting on the deck at the lodge, talking to a guest. He was facing south and the guest was facing north. The gentleman noticed the reflection of a fire in a window of the lodge.

"There's a fire over there," he said.

Chris looked and nonchalantly agreed, "Yeah, there's a fire there."

"There's a building on fire," the guest stated, his voice rising in alarm.

Chris thought to himself, *What do you expect? You're in a fire zone*. And then all of a sudden, the significance of what the man was saying struck him. "What? There's a fire there!" Chris exclaimed. He offered to drive over, see what was going on and report back if there was cause for concern.

The blaze was at the BC Wildfire camp and when Chris got there, he found the facility on high alert. All the sirens were going off. Apparently some hoses had caught fire. The camp was in the process of being shut down and the hoses had to be dried before they could be stowed away. For that purpose they were hung on a tower. Something went wrong with the propane dryer, or perhaps someone had cranked it up too high. In any case, the hoses ignited, and the tower itself burst into flame. Most of the firefighters had already left the camp; army personnel were in charge of closing the place and they didn't know much about firefighting. A great melee ensued. People were running around, yelling and screaming, trying to figure out what to do. One man tripped in all the confusion and broke his leg.

When Kurt heard the story, he thought, "Now the fire camp is on fire. Does it ever end?" The fire was successfully contained, and the camp was decommissioned without further ado. But in a way, it didn't end. As Kurt pointed out, the effects hung on: "When people come back from a war, they have battle fatigue and trauma. It's a lack of sleep and really a lack of control. You're trained to have the control, to do unbelievable things. You don't realize how bad things were until after. Reality kicks in. I relate it to childbirth. You have nine months of prep, that day comes, and hopefully after a few hours that part is over. That's what I thought the fire should be. I had days of prep and then the fire is gonna come and I'm going to beat it or lose. And it's over. But it came, it came, it came, it came, it came … "

I think almost everyone who went through the fires of 2017 was changed by the experience. But for Kurt and other people in Riske Creek, the feelings they had were especially intense because, in that community, the fires attacked repeatedly over a period of two months. As Kurt said, "There was no stop and finish. So I never had a chance to deal with it being over."

CHAPTER 16

YOU'RE GOING TO DIE, YOU NEED TO GET OUT OF HERE!

BRYAN POFFENROTH, A RANCHER, WAS one of the legendary locals in Riske Creek who did much to save his community from destruction. I phoned several times and left messages but received no replies. I wondered if for some reason he didn't want to talk about his experiences, but Kurt Van Ember encouraged me to stop in at his place. It was only a couple of kilometres south of the Chilcotin Lodge, on the road toward Farwell Canyon, and legendary in its own right. Back in the late 1800s, Fred Becher established a thriving hotel, trading post, saloon and ranch there. The hotel, saloon and store are gone but a dam and a tract of grassland still commemorate the pioneer's name and the ranch remains in operation.

Fortunately, Bryan was at home when Gordon and I knocked at his door. We introduced ourselves and explained that we were researching a book about the fires. Bryan welcomed us into a comfortable house with a large kitchen. We also met his wife, Raylene, and his young granddaughter. Then he offered us coffee as we settled in to hear his story. He didn't seem at all reluctant to tell us what had happened.

Bryan said that on July 7, he was haying when Raylene, who was in Williams Lake, called him from the Walmart parking lot. She sent him pictures of the fires erupting in every direction, all around the town. Bryan himself could see the smoke in the west. And when his son-in-law, Evan Fuller, came home to help pack up the last of the hay bales, he said, "When you look up to the west, it's burning good."

The BC Wildfire Service did not send any of its own crews to Riske Creek on that first day, it was so occupied with trouble elsewhere. However, it did not abandon the village either. Someone from the service phoned Lee Todd with the news: "We've got a fire at Raven Lake. We have no resources. Could you see what you could do with it?" The fire was just behind the Ilnicki Ranch, on the western side of Riske Creek. Lee was born and raised on that ranch; his father had sold it to Tom Ilnicki when Lee was fourteen. So he was familiar with the area and was friends with many of the residents, including Bryan, who liked and respected him.

"I have a construction company with thirty or forty machines," Lee told me. "I've always got a few Cats around. That's how it started. We sent a couple of my crew with a couple of my low-beds and we got some equipment out there. But it was almost a case of being too little, too late. We had no backup. The ministry was totally occupied by a couple of fires close to town."

Lee also flew to Riske Creek himself in his helicopter. He landed at the Ilnicki Ranch, now owned by Raylene's brother Fred, and saw Raylene. She had returned from town and happened to be there. Lee told her he needed a Cat operator to build a guard and asked her whether Bryan was around. When she called him at home, he came over immediately.

"We started working on it," Bryan told me, "getting guards in." The plan was to protect the ranch with a hundred-foot-wide firebreak. Lee made certain no one was taking crazy chances. Bryan asked him for a "flyby" to find out how things looked from the air—to see if they were okay. "He'd tell you if you were good or not. Everybody who knows him is willing to trust him, that's for sure."

In a way, this was an old-fashioned saga: a small group held together by strong ties of kinship and neighbourliness fights for its very existence. I could imagine it as a classic Akira Kurosawa epic. But it wasn't a simple pick-and-shovel operation. These people had modern assets, including Lee's aerial reconnaissance to guide their efforts and some of the heavy equipment they needed to defend against a powerful force of nature.

"We were thinking we were going to get it slowed down," Bryan said, "and all of a sudden we had to leave that guard because the fire came too fast. It started kicking our butt. We were trying to guard it too close for the conditions but we just didn't realize it." Big embers were blowing off the fir trees, jumping past the guard, and igniting small fires that came ever closer to the main ranch house. "At one point," Bryan said, "it got a little scary, a little crazy. We had to evac out of there."

Bryan Poffenroth, who owns the Becher Ranch, fought the Riske Creek fires all summer long.

The fire engulfed a garage and reached the house but didn't overtake it. This was a win for Bryan, Evan and Raylene's nephews, who'd worked hard all night. Nevertheless, they did not manage to prevent the fire from moving east. The next day, Bryan got on a Cat again in an attempt to create another guard and stop the voracious beast in its tracks. By 10:00 or 11:00 in the evening, when the break was almost complete, he parked the Cat and a skidder beside it, where he thought they would be safe. On July 9 at 4:00 in the morning, he left home with a young assistant to finish the job. Driving up a back road toward the guard, they ran into flames. Once again, the fire had outfoxed them. "We had to wait until it was light enough for Lee to be in the air to fly over, let us know if our machines were still standing and if it was okay to go in there. Lee said, 'Yeah, get in and out. It might get warm but get in and out.'" They rescued their equipment and escaped without injury but had to abandon the guard. "We worked for three days and three nights," Lee said. "I flew, like, oh man, from 4:00 in the morning in the dark until 10:00 at night. We saved every place at Riske Creek. We did what we could."

In those days early in July, several families sought refuge at the Poffenroths' ranch, which sits on the eastern side of Riske Creek and for the time being was out of danger. "I think for a couple nights we had thirty-five people and almost forty head of horses—and rabbits, dogs, cats. We should have done a group photo with all the people, definitely a gaggle of them," Bryan said.

"The one night up at my brother-in-law's, we were trying to be as safe as possible but we didn't really have a good head count, so I was kind of worried. When I got home, I made everyone write their names on that." He gestured to the roster, which he still had. "If you left, if you weren't staying here anymore, I'd cross your name off. If you were staying here, we wanted to know." He had an evacuation plan if the house was threatened. "We figured if it got that bad, we'd take a case of beer and go sit in the creek and watch it burn. There was nothing else you could do." But if it came to it, Bryan wanted to know how many people he needed to count.

The CRD issued an evacuation order for Riske Creek on July 9 at 12:30 in the afternoon, and just afterward members of a search and rescue team came to the ranch to give the Poffenroths the news. The children who were staying at the ranch would have to go. "The guys who had kids here, we got hold of them on radios so they could come and say goodbye," Raylene remembered. A tearful scene ensued. Families were separated and no one knew when the evacuees would be able to come back.

"It was hard when everybody left," Bryan said. "When everybody was here, we knew where they were. We knew they were safe because of the green hayfields and the creek. It wasn't coming there. When everybody was scattered, it got tough."

"The whole province could have lit up," was Raylene's view.

Once the children were gone, neither search and rescue nor the RCMP tried to persuade Bryan or Raylene to leave. "We didn't get harassed much at all," Bryan said. But there were confrontations. Lee Todd told me how one old rancher reacted when a young constable from Regina tried for a second time to get him to go. "The Mounties come, dark in the evening. He hollered, 'The next time you come rattling my gate, I'll be shooting! Now get the fuck off my ranch.' And he meant it."

"I'm sure he did," I said. I didn't envy that young constable at all.

THE BC WILDFIRE SERVICE FINALLY dispatched their crews to help on July 11. They had their own ways of doing things, not always to the liking of veterans like Lee: "They have to have a safety meeting at 9:00 in the morning. So then it's over at 10:30, 11:00, right when the most dangerous part of the afternoon is coming for the fire. That goes on steady. It's criminal, is what it is." Lee also objected when the local incident commander called an end to night shifts. "For forty-five years, I've been fighting fire and that's the safe time—at night. The humidity is up, the temperature is down. The whole fire, 99.99 percent of the time, goes to sleep. It doesn't go out, but it goes to sleep, goes down to cat 1 or less, compared to cat 4 or 5 in the afternoon. 'All right,' I said, 'If you don't keep a night crew up there, it'll jump your guard.'"

Raylene Poffenroth crossed checkpoints to warn her friends when fire came.

Sure enough, the fire crossed the line of defence. It would travel all the way south to the Gang Ranch and hit Riske Creek multiple times. "It burnt one of the twelve houses that I saved," Lee said. Could the fire have been contained if spark chasers had been on duty at night?

In their report, George Abbott and Maureen Chapman weighed in on the subject of night shifts: "We learned that the late-night to mid morning period was generally when wildfires were least active. Leveraging those windows to their fullest could enable BCWS teams to get ahead of wildfires."[49] The BC Wildfire Service already does conduct some night operations; maybe it should consider this option more frequently. In any case, there was nothing

Lee could do. The fire was out of his hands and he went back to Williams Lake.

By July 12, the fire was threatening Mike and Connie Jasper, who lived about ten kilometres north of the Chilcotin Lodge. They had lost power and were using five-gallon buckets to haul water from a creek—all they had to douse the roof of their house when embers landed on it. Bryan talked to one of the red shirts to see if he would provide a bladder for the Jaspers. He said, "We don't have any to spare."

"But they're pretty good guys," Bryan said. One of the firefighters asked Bryan if he could see the house where Bryan was thinking of installing a bladder. "He followed me for a drive. Mike and a cowboy, Jeff Taylor, down the road, had put in a hand guard on this hillside and the forestry guy took a look at it and said, 'Holy smokes. This is exactly what they teach us.' I told him, 'We're not dummies. We've been fighting fires here for thirty years or so, as long as we've been here. It lights up every couple of years.' And then we went to the house and I showed him what I wanted to do."

As they were driving away, they passed a ditch where a couple of bladders were being used to protect the garbage dump. The firefighter said, "I'm going to be gone for a bit. If one of these isn't here when I get back, that won't be a big deal." Bryan's son-in-law took the cue, emptied one of the bladders, loaded it on his truck, and set it up at Mike and Connie's house. "The guys with boots on the ground are pretty darn good," Bryan said.

Bryan also got help from one of the Alkali Lake crews who were working in the area. He was looking for small nozzles that would fit a 5/8-inch hose when one of the Alkali firefighters asked him if he needed anything. He mentioned nozzles. The firefighter said, "It's the one thing I can't give up. I've told my guys they've got to guard them. We're short and we're just not going to get any more right now." Bryan said, "No problem." Then he got into his truck and as he was driving along, he noticed a small nozzle sitting on the console. He told me, "One of the young guys, I don't know which one, had put it in my pickup."

When Bryan needed more fire hoses and the local stores in Williams Lake were completely sold out, a friend of his in Dawson

Creek offered to help. "He got us two thousand feet," Bryan said. "The funny thing is, it couldn't get through the roadblocks, and so we had a guy get it to a ranch north of here. They put it in one of the ranch pickups and came 'round the back roads and delivered it to us. You got two thousand feet of fire hose in your truck heading toward a fire, they should let it through."

I couldn't see how turning the truck back made sense either. I knew the police were worried about theft—with reason, as it turned out—but I still thought they should have been able to establish that the delivery was legit. I heard complaints about the checkpoints repeated over and over again. Here, for instance, is Lee Todd telling me about a roadblock at Tatla Lake: "I ran one of my own low-beds as hard as I could and I was in a helicopter. I was twenty hours a day for two months. I'd be going out to Kleena Kleene, taking some machines out there. I've got a D-8 on with the blade sideways and I'm sixteen feet wide. Well this one cop wasn't gonna let me through. 'Where do you think I'm going, dear? I've got a pilot car and a D-8 on the back. I'm going to help fight the fire.' I was sent out, but I'd left my little fire pass in my pickup. I mean, do you or do you not have a little bit of common sense? Gord Chipman come roaring up with a pickup. We were there a half an hour and she wasn't going to let us through until he came along. Eight out of ten Mounties would've said, 'Well, obviously get to the goddamn fire.' Oh man alive, I was just about ready to drive over her cruiser—and I would have. Like, give your head a shake."

IN CHAPTER 14, I DESCRIBED how the blaze attacking Riske Creek on July 15 was so aggressive it destroyed Ed McDonald's cars at Chilcotin Towing. That day the BC Wildfire Service decided a tactical retreat of its ground crews and contractors was in order. Three logging companies—Eldorado Log Hauling, Hytest Timber and Peterson Contracting—had arrived at the Poffenroths' ranch to help and they were supposed to leave too. However, as Raylene told me, "they didn't listen to what the ministry said. They just kept working of their own accord." Bryan already had a couple of bladders and those two thousand feet of hose. Thanks to the logging companies, two D-8 Caterpillars and two water trucks, one quite large, were also on hand. Bryan intended to stand his ground. However, he told me some of the BC Wildfire Service crews were on a sidewalk outside his house. A firefighter was "hollering and screaming, 'You're going to die. You need to get out of here. The fire's coming.'" He pointed to flames shooting up behind Becher's Prairie, a grassland area about seven kilometres to the northwest.

"That fire, right there? We're ready for that," Cory Dyck, the president of Hytest Timber, responded.

"And we were ready," Bryan told me, laughing at the memory. "We had a good setup up with two bladders, two pumps—three pumps actually—with a spare one just in case. And it's funny: when we set up, my son-in-law, Evan, stuck the last pump in the pool. Raylene said, 'If you're going to go to that one, it better be trying to take the house because it'll wreck my liners.' They ended up not using it." On the July 16, the fire started bearing down on the Poffenroths' ranch. But Bryan had braced himself for it. And between 3:00 and 4:00 in afternoon, the BC Wildfire Service sent in three helicopters to pick up water from behind Becher's Dam, a couple of kilometres west of Riske Creek, and bucket it across the road from the ranch until nightfall. They helped to beat the fire back.

The Wildfire Service ground crews and contractors soon returned. And then, for a while, the fire's fierce energy seemed to dissipate. During the lull, Bryan and Raylene went to their son's wedding in Alberta. "The timing couldn't have been better," Bryan said with a grin. They left all the hoses and pumps set up, just in case, and had a couple of fellows staying at the house to keep watch. When they returned from their trip, they discovered that the two caretakers had absconded with their hoses and pumps. "We had to beg, borrow and steal and get it set up again," Bryan said regretfully.

On Friday, August 11, the BC Wildfire Update on the Hanceville-Riske Creek fire reported that some hot spots had been actioned, a small "excursion" spotted the previous day was contained, and the road was continuing to act as a guard. None of this sounded alarming. That day, a structural protection specialist came to Riske Creek to see which houses needed attention. Bryan drove him around in his pickup to introduce him to some of the local homeowners. They pulled up to one lady's place, a small cabin tucked in behind some trees. "Is that even worth protecting?" the specialist asked. "It is to her," Bryan said. I could tell how shocked he was by this nonchalant attitude. Later Bryan found out that the expert had told a number of residents, "You're good, you don't need anything."

But that was not to be the case. The Friday update also predicted that the next day a cold front, never a welcome development, would be passing through: 20 kph winds gusting up to 40 kph were expected. Folks with respiratory concerns were advised to stay indoors; smoke might get heavy at times because winds, fire behaviour and temperatures were going to change.

As forecast, on Saturday, August 12, the winds came back and breathed new life into the fire. Most of the activity was to the west of the Poffenroths' property along Highway 20. Raylene drove over to see Leland Jasper, about fifteen kilometres away, then dropped in her on her friends Shelly and Hugh Loring. Just two days before, the couple, who had been sheltering their ponies at the Poffenroths', had felt safe enough to bring them home. "It was coming," Raylene recalled. "So I stopped at Shelly's and then I came back here to hook up my trailer to get her ponies." Raylene encountered a military roadblock on her way back with the trailer but she just kept going. She told the soldiers, "I'm not stopping. I'm going to evacuate the people." Raylene recalled her frustration with the blockade. Although there were two hundred soldiers stationed at Riske Creek, none of them were helping her get the word out to the other residents. "All they did was shut the road down."

Rita Rankin, who lived just west of the Lorings, was one of the homeowners who had been told she didn't need any structural protection. Bryan took her a cube of water anyway—a good thing too, because the fire arrived at her place about 3:00 in the afternoon. "It was roaring just like a 747 was coming," Bryan said. While he and a couple of other neighbours were helping Rita, another group was down at Pat Jasper's house, also on Highway 20 and also under attack. "We knew something was going on but we didn't know how bad it was," Bryan said. He had been at Pat's house earlier in the day wetting things down. When the wind seemed to abate, he'd figured everything was well in hand. He left a tank of water on the property and went home while a few other individuals stayed to keep watch.

Noreen and Ed McDonald were among the people who tried to help Pat Jasper, one of Noreen's brothers. But soon it was apparent that no one could do a thing. Raylene told me that the BC Wildfire Service had called for a bomber, but it couldn't be used. There was too much smoke. "It went dark here, blocked the sun right out," she said. A crowd stood on the road and watched the Jaspers' place burn. The house Noreen's grandfather had built, where she had grown up, was on the same property. Made of fir boards and insulated with sawdust, it exploded. "The embers were coming down around us and starting fires in the ditch on the side of the road," Noreen said. "When you breathed, you breathed in the heat. There was such wind blowing, it was scary. Ed finally said, 'I'm going,' because he had a tow truck with bikes and skidoos on it. We thought, 'If we stay here, our vehicles are going to burn.' The smoke was so thick, when we pulled out of the driveway to go onto Highway 20, you could not see if there was somebody coming. You just had to take a chance, pull out and hope."

When Bryan talked about Pat Jasper's house, he shook his head. "That shouldn't have happened," he said.

The Poffenroths' house survived, though Bryan lost grazing land and a couple of kilometres of fence. Ten cows and a bull went missing. "The government paid a little bit for them. It all helps, but not what they were worth." He had to get some emergency aid to feed his cows and he cut back the size of his herd. Months after the fire ended, he was still seeing the effects on his animals. "I shot one cow here in December. She kept getting snottier and snottier and drooling and couldn't breathe. I should've gotten an autopsy. I should have got them to open her up and look at her lungs. I also had one slip a calf [miscarry] and I don't know whether the fire had something to do with it."

As we were leaving, Bryan gave me a large map of the Hanceville–Riske Creek complex, dated August 26, 2017. A big, pink expanse in the middle shows the area burned. It extends about seventy kilometres from the northernmost point of the burn to the southernmost point and seventy kilometres from the western to the eastern border. The fire had ripped through a total of 232,444 hectares.

The map paints a grim picture, but it doesn't tell everything. It doesn't reveal what a strong community can do and what people who care for each other can accomplish when they pull together. Mike and Connie's place, Rita's house, the Lorings' home—all were saved. "We fought tooth and nail," Raylene said. If the local residents had gone when the evacuation order was issued, she maintained, "There wouldn't have been anything left."

CLINTON

CHAPTER 17

OH FIRE JUST TAKE ME

WHEN I FIRST MET RYAN Lake, he was wearing a black Stetson with a long feather stuck in the hat band. He was a sharp-featured man with a big grin and gleaming blue eyes. There was something a bit mischievous about him. You could see it in the way he had nicknames for things. Vancouver was always Metro Madness or M/M and his own home in Clinton was Whiskey Wells or W/W. And messages, whether sent by email or phone, were smoke signals.

One slushy day in December 2017, Ryan drove down to Vancouver to visit his grandson. I invited him to lunch at my house and we had a long conversation over bagels and lox.

Ryan is an ex-fishing-boat captain who had trained in marine firefighting. "But fighting bush fires is not the same. You're not surrounded by the extinguishing element, the ocean," he pointed out. However, he had attended a forum in Clinton the previous year and learned how to reduce the risks on his property. He replaced his shake roof with a metal one and disposed of a pile of building debris. He got busy with a chainsaw and cut off the lower two metres of branches from his trees. And he found a creative solution for trimming the grass on his one-acre property. "Got my neighbour's horses in actually. They did a great job. They munched everything down. He got free feed and it sure beats running around with a lawnmower. They even edged *over* the barbed wire!"

On July 14, the Elephant Hill fire was approaching the Bonaparte River, which rises on a high plateau northeast of Kamloops, then travels west and south to join the Thompson River at

Ashcroft. Gusty winds were forecast and the Thompson-Nicola Regional District (TNRD) issued an evacuation alert for Clinton at 5:00 p.m. Ryan ramped up his efforts. In addition to seventeen sprinklers and hundreds of feet of hose, he had four-outlet manifolds on all the outside taps that allowed them to be connected to several hoses at once. He'd also had twelve sprinkler stands welded up and placed them around the property where they had the reach he needed to cover everything. Remarkably, he did all this without his left leg, which had been amputated years earlier. Not only that, he was frequently unable to use his prosthesis. Because of all the hard work he'd been doing, he rubbed the skin on his stump raw. He had to tend to his hoses while on crutches, something he hadn't counted on.

A couple of Ryan's neighbours helped him position the sprinklers. "Some of them were up on rock piles; I'd be really challenged to get up there at all, never mind with a steel sprinkler stand. It took three of us to get it so the spray patterns were intercepting, overlapping, not spraying the road or cars going by. We had brass Quick Connects on all the sprinklers and on all the hoses, so if I needed to run an extra-long hose all the way down to the front gate, I could join hoses and get down there. That saved a lot of hassle. I can't kneel. I've got to lie down. Working with pliers [needed for older threaded connectors] would really have made it difficult."

As Ryan prepared his own place for the fire, he also watched the BC Wildfire Service at work: "You know, when you're driving north from Clinton, you go up that big hill and there's a rest stop on the right. That was crowded with the media and local people, some of whom had ranches down there where it was burning. The police were up there and the guys that spot and direct the choppers with their radios. Quite a party going on up there.

"Suddenly there were these huge balls of flame, every tree in the area would candle. It was a holocaust—an awesome force." At first, Ryan didn't understand why the trees were exploding, but he stood next to one of the air traffic controllers who told him what was going on. They were lighting a backfire from choppers to get rid of the fuel that the fire was advancing on—lobbing little balls about the size of ping-pong balls. "They ignite on contact with the ground," the air traffic controller explained. "They're safe on board but once they're away, they get like a weapon, cocked, and when they hit, poof."

Later when I talked to Glen Burgess, who was an incident commander in Williams Lake and on the Elephant Hill fire, he explained that the balls were filled with potassium permanganate powder and were completely inert and safe on their own. The machine that casts them overboard injects them with glycol (antifreeze). How much glycol it injects into the balls determines how quickly they explode and makes it possible to delay ignition until the ball hits the ground. "And what if you find one of these balls on the ground?" I asked him. "How safe are they to handle?" "They're not like unexploded ordnance," he assured me. "They probably didn't get injected."

ON JULY 29, THE ELEPHANT HILL fire crossed the Bonaparte River, a major milestone. At 4:30 in the afternoon, the evacuation alert for Clinton was upgraded to an order. Ryan said, "The day the order came down was my seventieth birthday so I had all these people phoning, a) to wish me happy birthday, and b) to make sure I wasn't a crispy critter yet. All these people, starting with the police, telling me I had to get out. I told them, 'Sorry.' They were really civilized. They didn't try to muscle me. They asked me who my dentist was. I know the drill. I've been dead several times and I'm seventy years old so I've had good innings. I wasn't too worried." The RCMP made one last attempt at 4:00 a.m. on July 30 to persuade Ryan to go. But his mind was made up.

Ryan's small A-frame home is in an enclave of seven households on Valley Drive, just north of the town proper. Five of the neighbours also elected to stay, so Ryan had company—a good thing, as it turned out.

Another good thing was the structural protection crew, which arrived shortly after the evacuation order was issued. "I hadn't realized there was such a thing—six of them from all over the place," Ryan said. "One was from Big White, another from Nelson. They were really professional, and boy did they move. They swept over us like a swarm of locusts." They attached sprinklers over the patio, the porch and the roof. "The structural crew could get to the peak of the house," Ryan said. "I can't do that anymore." The members of the team set up a couple of swimming-pool-sized bladders leading to the seven houses on Valley Drive. "When the bladders were depleted, they had tanker trucks refilling them. To protect my porch from embers, they stapled this heavy-duty plastic right around the whole front, which was facing in the direction of the fire. They did more work in a few hours than I could have done in a week—zip, zip, zip, zip. They were well trained, well coordinated. One of them was a British doctor; she can't work here, so she works with a volunteer fire department. I think she was one of the ones from Big White.

"They were very good with suggestions, like what you want is a mist, a mist around anything that's flammable. You don't have to have a hose on it, just a mist. My sprinklers were adjustable and some of the small ones I got, you can send out a mist. I've got them under the porch, so if this happens again, I'll be sprinkling under the porch where I've got a bunch of flammable stuff stored." The structural protection unit focused on people's homes. The team didn't do anything about Ryan's pumphouse or shop, or the shed in which he kept his tools. Nor did the crew attempt to provide any way of putting out spot fires that might start on the grounds around his house. The system Ryan had set up could do those things, although it was slow. It was running off a forty-year-old well and he could only operate a sprinkler for a maximum of an hour. Then he had to turn it off, wait an hour for the water to come back up, and start again. It was not an automatic system. He couldn't just leave it on. Keeping his outbuildings wetted down required constant attention, day and night. And the fact that Ryan had to hobble around on crutches, dragging heavy hoses over uneven ground, made the task even harder.

No wonder, at one point in the middle of the night, while repositioning a hose, Ryan tripped over something. "I'm lying on my back, looking at the sky." He noticed the moon and thought, *Oh fire just take me!* And then he had another thought: *If I lie here any longer, somebody is going to look out their window, the girls up at the corner, their place overlooks ours, they'll look down and pan-*

ic, they'll think I've had a heart attack or something. "So I got myself up and carried on dragging hoses around on crutches."

Ryan wasn't expecting to fall, but he wasn't surprised either. He has bursitis in both hips, an inflammatory condition probably related to his use of a prosthesis. "It's just a fact of life that if you have an above-the-knee prosthesis, walking is the equivalent of my able-bodied twin running up stairs," he said. His remaining knee was pretty much worn out, and by over-using crutches he had also developed rotator cuff injuries in his shoulders and elbow. "Sometimes I forget that I'm not nineteen and ten feet tall. I overdid it," Ryan said.

Ryan Lake stayed in Clinton to protect his house, even when an evacuation order was issued for the town.

Ryan's worst moment was when a finger of fire came down the Hart Ridge to the south. He had seen a huge glow behind the ridge for some time. A friend in Texas had been emailing him maps twice a day and by poring over the latest installment he had an idea about how large that fire was. Behind the finger was a monstrous hand and arm. He consulted links on websites that the Village of Clinton and Environment Canada provided, so he knew the winds were pushing the fire toward him.

"And then," Ryan said, "at the critical moment, when the elephant came charging down the hill at us, the pump air-locked. I couldn't prime it the way it should be primed. The bolts were frozen in place. I had to take a hose off the pump, pour the water in backward, run the pump and keep pouring water in. When it started spluttering out all over me, the pump was catching and I'd shove the hose back with one hand while I reached for a wrench with the other."

While Ryan was in the middle of this onerous operation, he was covered in mud and soaking wet. The fire was less than a kilometre away and blowing embers toward his place. Even though he was buffered by a green field across the road in front of his property, Ryan knew that if the wind picked up, the embers could pass over the field and start blazes long before the main body of the fire stormed in. Ryan was in a race with the elephant.

Then the two gals from next door, Morgyn and Kim Bosche, arrived for a barbeque. The three of them had developed quite a bond during the fires and often had dinner together. Morgyn and Kim were in their party clothes, but they waded in to help Ryan nonetheless. "With the pump out of action you can't use a hose to prime the pump because the hoses are all dependent on the pump," Ryan explained. But he was prepared for this; he had water stored in horse buckets. Morgyn and Kim carted them around and helped Ryan pour the water into the pump. Soon they too were drenched, but the trio finally managed to get the pump going. "It took six hands to do it," Ryan said.

"I'm one of the luckiest people you'll ever meet. I have guardian angels," Ryan said. The wind shifted and blew the fire back up the ridge toward the Skookumhorse Ranch. "From that point on, it was like *M*A*S*H*," Ryan recalled. "There were three or four choppers in the air at any given time dropping fire retardant. Ground crews with heavy equipment were cutting firebreaks. Choppers were everywhere. They swoop in and pull out almost like a strafing run. I was envious. One of my life's ambitions was to be a chopper pilot; I'll never do it now. I was down on the ground thinking I could trade places with any of those guys. They all had a technique. You could tell by the different colours of choppers. They all had a different way of going into it. One guy would sort of hover, then he'd disappear into the smoke, and then he'd come straight back up and away. Pretty cool. Others would take a flying run at it and take off, because they were dangling buckets."

The danger passed for the residents along Valley Drive, though the Skookumhorse Ranch was under serious threat for some time. Ryan maintained his routine of damping down the place. He was able to keep supplied with whiskey and other essentials through Jinwoo Kim at Budget Foods in Clinton. Ryan couldn't visit the store, as that would have violated the evacuation order. So the two arranged to meet at a checkpoint where Jin could deliver what Ryan needed.

At 11:30 a.m. on August 15, the evacuation order was downgraded to an alert and life gradually returned to normal.

RYAN AND I TALKED FOR a couple of hours. Then the bagels were gone, the tea had grown cold and Ryan had other stops to make. I told him I was interested in visiting Clinton and meeting people affected by the fires. We promised to stay in touch.

Over the next couple of months, the smoke signals, as Ryan would refer to them, went back and forth, and then Gordon and I drove north to meet some of the people I had been talking to on the phone.

Ryan invited us to have tea with his helpful neighbours, Morgyn and Kim, and their dog, Darwin. In his homey and somewhat cluttered house, I taped our conversation. (Later, when I used Trint software to automatically create a transcript, I was amused to see Darwin's comments rendered as "Ow ow ow ow ow ow ow ow ow ow ow ow ow ow ow!") Morgan, who did most of the talking, is a lanky woman with straight blond hair. Kim is rounder and shorter, with curly red hair. After moving to Clinton from Abbotsford early in 2016, the couple had married in June of that year. They own an IT business and Ryan described them as "tech savvy," probably a valuable skill to have during the fires.

Though Morgyn and Kim were newcomers to the Cariboo, they seemed to be as ingenious as some of the old-timers. They laid in stores of extra food and respiratory masks designed for smoke. They got a generator so they could survive power outages. "If you're on well water, you have to provide your own power. If the utility power goes down and you don't have a backup battery or a sufficiently large generator, you're left helpless," Morgyn said. Both women are gardeners, so they were equipped with an irrigation system that they rejigged to sprinkle the property.

They also turned their truck into a small tender with fermentation vessels they had for making their own beer and wine. They had a dozen "carboys"—rigid containers with narrow necks that each held seventy to ninety litres—as well as a few open fermentation pails. They filled these with water and put them under the truck canopy. Then they took a pump from their greenhouse, connected it to an inverter and a battery to give it power, and attached a couple of hoses. When they hit the switch they could start pumping. While they couldn't put out a huge fire, they had a mobile system that could extinguish small blazes in their enclave *before* they turned into major problems. "We configured the truck before we realized that the fire department was coming to put sprinklers everywhere. We were able to tell them to put more sprinklers on the other side of the street because we were covered. We were really confident that we were in a good position if the fire came."

Morgyn has a thirst for information. A ham operator, she listened to the local fire department on one channel and the air crew on another. "What I inferred from what I heard on the radio was that there were some major communication lines that run though the Clinton area. After the fire jumped the highway down south and it was approaching those lines, they doubled down on their efforts to make sure it was put out here. Their attack became relentless. Where we saw a fire, they sent the helicopters in with water—three water drops every two minutes—constantly, until that fire was put out."

While the fire was playing over the Hart Ridge, Morgyn had a security camera pointed at it all the times. It took time lapse photos at regular intervals; this allowed her to see what was changing. At one point, she saw what looked like smoke, rather far away, in some of the images. Thinking she had discovered a new fire, she reported it. But when she examined the pictures more carefully, she realized that she was looking at dust, not smoke. Operators using excavators and other heavy equipment were creating a break to prevent the fire from spreading. It was nothing to be alarmed about.

Morgyn's meticulous observations made her a trusted source. During the fires, people had an insatiable demand for reports about their local area. This was true not just for the people who had stayed to protect their places, but also for those who had left and were watching helplessly from the sidelines. When Morgyn got on the roof and filmed herself giving a local newscast with the camera she had been using to survey the Hart Ridge, she got thousands of hits. She was filling a vacuum. Then CBC asked for reports. I was reminded of what Jordan Grier at the Chilco Ranch had said: "When there's nobody around to tell you that you can't, you would be surprised what you can do."

Kim and Morgyn Bosche created their own mini-tender to prepare for the worst.

AS RYAN GENEROUSLY DROVE WITH us to visit some of his neighbours and friends, I got comfortable enough to ask how he lost his leg. He didn't seem to mind the question. "It was at the May 3 Rodeo at Deadman's River, in 1997." On a whim, he'd decided to sign up for the bull riding contest. He drew a bull called Mr. Twister, a name he deserved. Ryan managed to stay on for eight seconds, enough time for him to win him the purse, but then he couldn't get off. When he finally slid down, the bull gored his leg. It couldn't be saved and he needed two years of rehab at G.F. Strong in Vancouver to recover and learn to walk again. "What an amazing story," I said. Ryan nodded.

The next day, we were having dinner together at Gold Mountain Restaurant, another Clinton institution. As we were passing around dishes to share, Ryan asked me whether I'd like to hear the real story about his leg. "Yes," I said. "It happened on May 3, 1997," he started. He was driving his car on the Cariboo Highway at Yale when he was hit. No one in the car was killed, but the car was so badly crumpled, it took the first responders two and a half hours to cut him out. A nurse who lived in Yale came to the scene, crawled into the car and cradled his head while the first responders worked to get the vehicle open. Ryan was taken to hospital and then to G.F. Strong, where he had two years of rehab. "I have another story about it too." His eyes twinkled. But this one sounded right to me.

I thought it was remarkable that Ryan could joke about his disability, that he made light of it. He didn't let it stop him. He never complained, just adapted and did what he could. It was typical, I thought, of many of the people I met. They accepted what life handed them and didn't ask for special favours, even when the going got rough. It's the Cariboo way.

CHAPTER 18

I HAD TO GO OVER THE MOUNTAIN

"WHAT ARE THEY GOING TO do about gas?" Jinwoo Kim was thinking. When the evacuation order was issued for Clinton on July 29, a couple of RCMP officers knocked on his door and told him it was time to leave. Jin, a big man with an infectious laugh, is the proprietor of Clinton Shell Gas and Budget Foods, right in the middle of Clinton. There are only two gas stations in town: the Petro-Canada and his. Jin knew that the owner of the other station had already shut down and left. He figured that many of the evacuees would want to fill up on their way out. Although he sent his wife and kids down to the coast for safety, he didn't think it made sense for him to go—not just yet anyway.

Jin lowered the price of gas to his cost and stayed open. "There are lots of people who need some financial help. At least that would give them enough to get out of town," Jin said to me. We were speaking in his office, up a precipitously steep set of stairs above his store. When Clinton was threatened, Jin posted on his Facebook page that he would wait until everybody else got out before leaving himself. Then an RCMP officer visited. He had realized that the firefight desperately needed Jin—the officers, the firefighters, everyone who was staying behind. He pleaded, "Can you please stay back? Can you stay to help?"

Thirty fire trucks had arrived in Clinton, some with empty tanks. More firefighters, police officers *and* more vehicles were on the way. Yes, there was a gas station in Cache Creek, but if the trucks went there to fill up, it would be an hour taken away from firefighting. Jin agreed to assist. The first week of the evacuation order, he was open every day for twenty hours. "'Cause they were setting up and all that and they always needed fuel. It was crazy. There was no planning.

"I'd be sleeping at home and they would come and knock. Everybody here knows where I live. They'd come at 3:00 in the morning. I'd come down to get them fuel. One morning they wanted me to fill up at 5:00 in the morning. I slept in that day and then two fire trucks came to my house. Twice they woke me up. Pretty intense." Jim reminded the drivers they owed him one. He said, "If you guys are leaving town, please take me with you. I've got no vehicle. My wife took the vehicles and all that. Don't you guys leave me behind."

During that first week, Jin was alone in the store; his employees had left. So he called his brother, Sang, and asked him to come to Clinton and help out. "I got him into town," Jin said. When I asked Jin how he got Sang past the checkpoints, he explained with a big grin: "I held the fuel. I said, 'If you want fuel, I need help.'" After Sang's arrival, things settled down slightly. Jin now had someone to spell him off and, since his brother had driven up, he also had access to a vehicle.

"The cops needed ice and energy drinks," Jin remembered. "They were working so hard, standing outside in the sun. It was hot and smoky. Coca-Cola, Pepsi and the ice company wouldn't let their drivers go into evacuation areas for the drivers' own safety. So I'd go meet them at Lone Butte or Cache Creek." Jin would grab the drinks and the ice and race back to Clinton, trying to get there before all the ice melted. Then he'd deliver the refreshments to the officers in their cars.

Clinton did not empty out after the evacuation was ordered. About four hundred firefighters came in. Three hundred of them were with the BC Wildfire Service; the rest were members of thirty different fire departments, large and small, located all over BC. Jim Rivett, who was mayor at the time, said that about a quarter of the town's 650 residents stayed behind also, some to support the firefighters and first responders by keeping restaurants and hotels up and running, and some, like Ryan Lake, to defend and protect their own places. All of these people needed food. If they left to pick up supplies, they would not be allowed back in. Those residents who wanted to stay turned to local outlets like Budget Foods to fill their needs. And although the Wildfire Service had fuel for its helicopters, the police and many of the ground crews relied on Jin for gasoline and diesel.

Jin had been having trouble getting in supplies even before the evacuation order was issued. Highways around Clinton were closed due to fires elsewhere and the delivery trucks couldn't get through. When he ran low on milk, bread and other necessities, he appealed to Jim, the mayor, for help. Jim told me on the phone, "Well, I intervened and spoke with the TNRD and CRD and explained that they didn't need to throw up all these roadblocks, and then it was, 'Well it's not us, it's the RCMP,' and the RCMP said, 'Well it's not us.' But by going to them and just kind of raising hell, they got the message and things got a little better."

"Once," Jin said, "I was out of fuel, empty. All the trucks were just waiting here. Thirty fire trucks waiting here. They're like, 'When are you getting fuel? When are you getting fuel?' Fire engines go through diesel like crazy." Normally in the summer a load of diesel would last him ten days or so. Now he needed two or three deliveries of diesel every week, as well as two tanker loads of gasoline. Despite the lineup at the pumps, the TNRD wouldn't approve delivery because it was too dangerous for the fuel to come in. Jin worked the phone. He talked to the mayor and to a good friend of his who knew the movers and shakers in the TNRD. Someone got wind of the situation and publicized the news on Facebook. Finally, Jin talked to the incident commander on the Elephant Hill fire at the time and got his approval to bring fuel in.

Sometimes red tape wasn't the problem, Mother Nature was. When fires broke out on both sides of Highway 97 to the south, it became simply too hazardous for drivers to make the journey. Jin decided he would pick up groceries himself by an alternative route. Highway 99 was also closed, but there was another option—a dirt road, very scenic and *very* hair-raising. "I had to go over the mountain on Pavilion," Jin explained. The road is a narrow track with an

CLINTON

Jinwoo Kim, owner of Clinton Shell Gas and Budget Foods, kept Clinton supplied with food and gas while the town was under order.

18 percent grade going up from Clinton to the summit and a 12 percent grade going down the other side. In places it's single-lane only, so if two vehicles meet, one of them has to back up to a wider spot. The road travels along tight switchbacks over steep cliffs, without any guardrails between the driver and the yawning chasm below. "But we needed to survive," Jin said. "I had to do the deal or the food truck wouldn't come at all." He did the trip three times with several friends in pickups. They met a grocery truck at the gas station at Fountain Flat Trading Post, about fifteen kilometres northeast of Lillooet. Jin told me, "The town people have been helping me to succeed in this business for twenty years, so I'm like, 'Okay, this is my time to pay back.'"

According to Jin, one of the consequences of the town being on order was that the residents "had absolutely no way to get to my store." This would be violating the terms of the order. Not only did he have to bring in food for them, but he also had to deliver it. Jin had a pass that allowed him to go back and forth; he would arrange to meet his customers at a checkpoint that had been set up in town. Then he'd explain to the police that he was from the Shell gas station, bringing in supplies. "They were pretty easy as long as people didn't wander around in town."

On the other hand, the firefighters were able to come into his store. This gave Jin a chance to get to know them, and he developed a deep respect for what they were doing. "Those ground crews, they work long hours. They'll get going at 7:00 in the morning, finish at 7:00. Then if it's crazy, they'll get called again at 10:00

and work all night. Come back and sleep for six hours and go out again. That's why I opened up my store for them. Yeah, I'm like, 'You guys are working your butts off to save my town. I can supply all the amenities, the snacks.' Sometimes I'd just give it to them. I said, 'Just take it.'" Jin told me that he also gave crews vitamin C. "It was cold some mornings and they're sleeping in a tent. They're very tired. I said, 'Just take it.'" I laughed when I heard this. "You were like their mom," I told him. He laughed too.

Jin managed the store with only his brother to help for about two weeks. When I remarked that it must have been exhausting running things during the fires, he said, "It was exhausting *and* stressful. But I had all the people behind me. People would say on a Facebook page, 'Thank you very much.' A kind word like that kept me going. While it's something that I wouldn't want to see again, when the hard time came, everybody got together and worked together. I met so many people and we became good friends. I still talk to them over Facebook. It was a good experience."

CHAPTER 19

OUR NEIGHBOURS WERE EVERYTHING TO US

August 2
4:15 AM

It is with the saddest of hearts that I have to tell you that our mountains are on fire all around us. This is due to a tragic back burn that went horribly wrong.

TERI-LYN DOUGHERTY WROTE THIS IN an impassioned Facebook post from behind the fire lines. When I read the piece some months later, I was struck by the depth of feeling she expressed. It is one thing to suffer a loss because of an unstoppable wildfire. I could see that it is quite another when the damage is the result of someone's miscalculation. I sent Teri-Lyn a message asking if she would be willing to share her story with me. Graciously, she agreed, and so on a snowy February morning in 2018, Gordon and I travelled south from Clinton to the Maiden Creek Ranch where Teri-Lyn lives with her husband, Kenny.

I'd done a little reading before I arrived. I knew that the ranch was historic. A few years ago, British Columbia presented the Doughertys with a Century Ranch Award commemorating the fact that their place has been in the same family for a hundred years. I also learned that Edward Dougherty, who immigrated to Canada from Ireland, obtained the land in 1862. This meant

Teri-Lyn and Kenny Dougherty, who live on the Maiden Creek
Ranch, dealt with prescribed back burns gone awry.

the ranch was actually over 150 years old. It existed even before British Columbia joined Confederation. When Edward arrived in the Cariboo, the gold rush was starting and tens of thousands of dreamers were pouring into the area hoping to strike it rich. Like so many others, he went to Barkerville, but after a while he decided to try something less chancy. He established a homestead on the Maiden Creek Ranch in the Bonaparte Valley and returned to Barkerville for a spell. There he met Elizabeth, who was working as a "hurdy-gurdy girl"—a woman who danced and played the hurdy-gurdy, a stringed instrument that sounded something like the bagpipes.[50] They married and Edward began ranching full-time. Because the famous Cariboo Wagon Road ran through his property, he also built a roadhouse to feed hungry travellers. The Doughertys had nine children and many descendants. In 2012, when the family held a reunion, three hundred people spanning seven generations showed up.[51]

Teri-Lyn and Kenny welcomed us warmly and we sat around a long dining room table, where they served us tea and homemade biscuits. A friendly orange cat settled on Gordon's lap. Kenny told us that the ranch extends over 650 acres and that his brother, Ray, looks after the business of raising and selling cattle (currently around 250 head). Kenny works at the sawmill in Chasm, but he is still a part owner of the ranch, along with his older brothers, Ray and Chuck, as well as their mom, Helene Cade. Kenny's nephew, Tyler, and a cousin all live at the ranch. Teri-Lyn grew up next door on a property that her father purchased. I find it remarkable that every day, Kenny walks where his great-grandfather walked before him and sees the sun rise and set over the same hills. In BC, you rarely meet people with such strong ties to the land on which they live. For the Doughertys especially, I think the fire must have felt like a profound violation.

I gazed out to a meadow, where I saw a couple of horses standing quietly, their dark brown coats lightly dusted with white flakes. The previous summer the scene was completely different. The Doughertys have a small, framed photograph that gives some idea of what they faced. Taken at night, it shows the living room picture window looking out on red flames leaping through the trees to the east of the house. Teri-Lyn also showed me a short video she made; as I watched it, I could hear the unmistakable crackling of fire. This February morning, the ranch was a world away from that, steeped in snowy silence.

Teri-Lyn and Kenny took turns telling their story, moving without interruption back and forth. Speaking of when the fire erupted in Ashcroft, Teri-Lyn said, "You can't imagine the wind that day. It was like a wind we've never experienced. It ripped the soffits out of the roof and they flew through the air. You could hear trees going down everywhere. The horses were just running; they didn't know what to do." They ran, but there was nowhere for them to run to.

Very soon after learning what had happened in Ashcroft, Kenny and Teri-Lyn began to prepare. They did everything they could to protect their house. They logged twelve nearby fir trees. They bought three hundred metres of fire hose and Kenny put a sprinkler on the roof. They moved both of their horses as well as three miniatures and a donkey, which they were safeguarding for friends, to safety in Clinton, as it was not on alert at the time. When the fire veered east toward Loon Lake, they relaxed a bit; they were not so directly in its path. But by July 14, the fire had swept through the community of Loon Lake. It stretched over forty thousand hectares, having quadrupled in size from just four days before. About ten kilometres to the south of the Maiden Creek Ranch, it was obviously quite intense and dangerous. When the TNRD issued an evacuation alert for Clinton, it also issued an alert for properties to the south, including the Doughertys' place. Teri-Lyn and Kenny wanted to move their animals even further away from the trouble, so Kenny's niece picked them up from Clinton and took them all to her place in Chase, two hundred kilometres to the east. Then Teri-Lyn said, "My fear level got to a point where I couldn't handle it anymore. I was just so terrified." The Doughertys also went to Chase.

On July 20, the TNRD downgraded the evacuation order to an alert and the Doughertys came home with their animals in tow. They were happy to be back, but they experienced sadness too. Teri-Lyn said, "Our uncle next door passed away and the day after the service, we were evacuated again. We had just enough time to have a funeral."

The reprieve from the Elephant Hill fire had lasted a little over a week; the TNRD's order to leave Clinton and the corridor along Highway 97 came on July 29. Because the fire was on the east side of the highway, the Doughertys believed that their house, on the west side, was not particularly at risk. They decided to stay put, although they did take the precaution of moving their animals again—into the care of their friends, the Woodburns, who had a ranch near 16 Mile House, south of the main fire activity (and whose story I told in chapter 5).

On August 1, the BC Wildfire Service tried to contain the Elephant Hill fire on its western flank by setting a controlled burn, intending it to consume fuel ahead of the fire and stop it in its tracks. Helicopters dropped incendiaries on a height of land just east of Highway 97, near Loon Lake Road. Unfortunately, Mother Nature did not co-operate. Extremely hot weather and erratic winds made the firefighters' work much more difficult. Controlled burns are a time-honoured technique for fighting wildfires. Before the days of water bombers and fire retardant, they were firefighters' main tool.

Burns continue to be used, but not without controversy. "That was the single biggest piece we were called to question on—the use of fire and the application of fire," Glen Burgess remembered. The issue invariably came up in the public information meetings the incident commander attended. "We burn every single day. On a small scale, the crews burn with their hand torches. It's ongoing. In fact, most days you won't even know we've done it because there's smoke from the fire and now there is a little bit more. Can't even tell the difference. On large fires, that's the only tool we have that is effective. We build a fireguard that's three metres wide or four metres wide not because we think that's going to stop the fire but that is a line for us to work from." Glen explained that when embers were blowing and starting spot fires a thousand metres or four thousand metres beyond a fire's main ignition zone, the guard wasn't going to stop anything. "We build a fireguard to give us what we call 'defensible space' to work from. We then put fire down to remove the fuel between the oncoming fire and our guard. This makes the guard hundreds of metres wide because we've removed all that burnable fuel."

Unfortunately, the prescribed burn set near the Loon Lake turnoff did not go according to plan. The fire jumped Highway 97 and threatened homes and ranches on the west side of the roadway. Kevin Skrepnek, a spokesperson for the BC Wildfire Service, stated that conditions appeared ideal until the winds shifted unexpectedly and dramatically. Greg Nyman, a neighbour of the Doughertys whose range was badly burned, disagreed. He said in a number of interviews with reporters that the wind shift was obvious before the fire was lit. While the Wildfire Service was relying on the winds blowing to the northeast, Nyman maintained they were in fact already blowing to the southwest when the burn started.[52]

Kenny said, "The day it jumped the highway was the scariest moment. We didn't see it jump but we were sure things weren't good down there because of the thick black smoke. The flames were coming over the top of the hill and all of a sudden there were five helicopters going like crazy. They were dipping into a little pond up here. Someone timed them; they had a three-minute turnaround." After the fire crossed the highway, a crew from Delta arrived as soon as they could to provide structural protection. They were the ones who told the Doughertys the fire had jumped. The firemen were plainly shell-shocked. They talked about a "fire tornado" four hundred feet tall that produced a powerful air current. It uprooted trees, which then toppled a couple of hydro poles. The wind was so strong the crew could barely get into their vehicles and close their doors before driving off, dragging their hoses.

Later, another structural firefighter from Vancouver Island showed Kenny the decals on his hat that had melted during the back burn. Kenny said, "He was sitting on the side of the highway when they were lighting all this. When it jumped the highway, he thought he was done for. He didn't have time to be scared or anything. He was just thinking, *This is it*. He crouched down beside his truck, turned on the hose attached to it, and got it spraying above him. The flames swept over his head, but he was lucky and he suffered no serious harm.

"When the fire crossed the highway, it tore up through this draw here," Teri-Lyn said, gesturing to the window. They moved some of their family treasures out of the house to a travel trailer, which they pulled to safety in the middle of a meadow. Then Barb Woodburn phoned. She was starting to get uneasy about her own place and wondered if Teri-Lyn wanted to move her horses. Teri-Lyn explained that the fire was approaching their ranch and they couldn't leave. She asked Barb to put the animals in a green pasture. There was nothing more she could do. As you know from the story in chapter 5, Barb's fears were justified. At one point, the fire that threatened the Doughertys turned south to her ranch but was repelled by their neighbours at 16 Mile. The Doughertys' horses survived.

"It was the most devastating thing we'd ever seen in our lives," Teri-Lyn recalled. "Knowing now that all our range, all of our cows were in jeopardy, not knowing whether we could get them out … we couldn't sleep. We stayed up watching the fire 'til the middle of the night. We went to the trailer and tried to sleep but couldn't. That's when I got up and let people know what was happening."

After these frightening events, structural protection crews started coming to the Doughertys' ranch every day. They came in the morning and didn't leave until 8:00 in the evening. Teri-Lyn recollected, "I can't say enough about them. Whether they were from Delta, or the Island, or down on the coast, they were all very kind. The ones that did the sprinklers here on the ranch were from Williams Lake."

The structural crews used the ranch as a base. They could keep an eye on things, but also fan out to other areas close-by that might need their help. Meanwhile, the fire threatening Maiden Creek was still burning. A group of firefighters started another controlled burn to create a fireguard on one corner of the property. Kenny said, "They had little torches and walked around. When it's lit by helicopter, it's really black, really burnt. Where they light it by hand, it's much smaller."

Curiously, the fire didn't burn itself out, but it didn't advance much either. It was like a cat, playing with them. Teri-Lyn said, "Every night, we would sit up and watch that fire come down the draw to our house. We were just coexisting with the fire right beside us. We were learning to take turns sleeping, to keep checking and seeing where the fire was and what the wind was doing."

After about two weeks of this, Kenny and Teri-Lyn were bone tired. At one point, Kenny talked to a firefighter, saying, "'It's still burning every night. We can't stay up forever watching it come to our house. You have to decide what you're going to do. Finish burning it or put the fire out for us.' The next day before dark, I don't know how many trucks rolled in. I asked one of the firemen 'Can I sleep tonight? You guys are here watching over things?' The fireman said, 'No. We're going to light it up. We're burning it.' I asked, 'Maybe I should put sprinklers on?' 'Oh yeah,' the fireman said, 'it wouldn't be a bad idea if you have some.'"

Kenny shook his head as he told me about it. "They were sending all the structure guys around to every house, making sure they were all set up for this. The day that they're actually lighting it, they don't bother saying anything. Once they finished lighting everything up, they packed up and left. They lit the whole side of the hill on fire."

That night, Teri-Lyn and Kenny sat out on their deck, once again on sentry duty. They had the sprinklers going but they could still feel the heat of the blaze, it was that close. "The structural guys were here for twelve or thirteen days straight and that day they didn't show up," Kenny recalled. I could see that he was still incredulous. "We were so exhausted," Teri Lyn said. Although they were trying desperately to stay awake, she simply had to have some sleep. She told Kenny she would go to bed for just a couple of hours. Then she went upstairs, looked out her window and thought: *Am I crazy? I can't go to bed. I have to do something to wake myself up again.* She felt she was in the most jeopardy she'd ever been in. She couldn't rest; she had to remain vigilant.

The cat-and-mouse game continued. Although the Doughertys had become somewhat inured to the fire, other people were still shocked by its proximity. Teri-Lyn remembers a young woman who had been evacuated from Sheridan Lake and who was driving down south on the highway. She came by to tell Teri-Lyn that a fire was threatening the ranch. "She was so kind and scared for us." Teri-Lyn explained that the Wildfire Service was aware of the blaze, but the young woman could call it in, if that would make her feel better. And the woman did call it in. Kenny said that every day, for about a week, at least one person would drop by to warn them about the flames on their property.

The evacuation order was finally lifted at the end of August. The crews returned home and Teri-Lyn and Kenny went away for a week to recuperate. Their house and animals had survived, although half of the Maiden Creek's rangeland was burned. Normally, it takes about three years to fully recover from such destruction. "For your mental health, you know you have to leave," Teri-Lyn said. "There's going to be a time that you have to say, 'It's going to be okay.'" But she found the adjustment difficult. "You see a little smoker here. You're thinking, if I leave, and there's nobody here to stop it and if the wind comes up again, who's going to stop it from going to your house? You can't leave for weeks on end and leave the sprinkler going on the roof. It could flood your house. The fall was really, really hard, wondering if there was going to be one bad day. There were little fires everywhere. You got used to seeing them. But you felt you had to be on high alert at all times. You couldn't just forget about it. You needed to pay attention. It could come back.

"When you stay, it comes with a mental cost. It really does," Teri-Lyn continued. "It's going to be with us forever. It will definitely change how we respond to things in the future." Teri-Lyn works with families who have young children. She became aware of how they were struggling to cope with the aftermath of the fires too. She said, "Not only did I carry a story with me, but all of the families in Cache Creek and Ashcroft had their own stories about how much it affected them and their children."

Kenny added, "You hear a helicopter now. You just instantly ... the helicopters. It was really tough. It's not the pilots. They're just doing their job. They're lighting the fires. But they're there saving you too. It's a real mixed feeling."

Despite everything, Teri-Lyn is optimistic for the long term. "I believe the land will come back," she told me. But in the short term, ranchers face significant challenges. In 2018, the BC government reported that over the previous summer, two thousand cattle had been lost. They either died or had to be euthanized. Fencing stretching over a length of eleven hundred kilometres had to be replaced. (This is just under the distance from the southern border of BC with the US to its northern border with Yukon.) Many ranches had to reduce the size of their herds because their rangeland was burned; losses to the ranching industry were estimated between $35 and $70 million. The rangeland is not expected to fully recover until 2020; obviously it will take time to rebuild and get back to 2016 revenue levels. The BC government promised to help; by 2018, it had paid about $8 million to 200 ranchers and farmers.[53]

The government also promised an investigation into the back burn at Clinton. A week after it was lit, Doug Donaldson, Minister of Forests, Lands and Natural Resource Operations, pledged that officials with the BC Wildfire Service would begin an inquiry to find out whether the back burn operation south of Clinton destroyed local cattle and property instead of protecting ranchers from the encroaching wildfire.[54] So far, no results have been published.

"We can't predict what would have happened if they hadn't done it," Teri-Lyn observed. "That's the tricky part. You have to believe that they are doing what's best in the moment. We were never faced with a moment like this. We can't go on experience, to know whether this was the correct thing to do or not. What we can say for certain is a lot of devastation was caused afterwards, because of that decision. We can drive you to the Lillooet Highway and show you the amount of devastation. If you're driving up the hill here and you know for a fact that it came from a back burn that jumped the highway, you get some sense of how much land was destroyed. What's going to be the future for the ranches because of it?"

Toward the end of our talk, Teri-Lyn said she had something she wanted to show us. She went away for a minute and returned with a large photograph of an old wooden gate. Kenny said it had been on their land for as long as he could remember. "That used to be a beautiful part of the ranch. We'd have weddings there. And now it's gone," Teri-Lyn said. "How do you replace that?"

The losses are undeniable. But Teri-Lyn and Kenny draw strength from their community. Teri-Lyn said, "I never wanted to come across as ungrateful for what everybody did to help and protect us. One bad decision doesn't negate all the work that was done afterwards. There were so many stories of bravery and what people were doing, trying to protect their neighbours' places and help their neighbours in any way that they could. Our neighbours were everything to us, whether they were helping us to bring over a water truck or sitting up with us at night to try and help us stay awake or give us supplies, when they could get us supplies. We're so fortunate to live with people we know so well."

PRESSY LAKE-70 MILE HOUSE-GREEN LAKE

CHAPTER 20

WHAT THE HELL WAS WRONG WITH US?

IN 1970, WHEN RAY PAULOKANGAS and his family drove up the Fraser Canyon and visited Pressy Lake, 110 kilometres northeast of Ashcroft, they immediately felt at home. The long, narrow lake fringed with forested hills looked a lot like Finland, which they had left just five years before. "We liked it," Ray said, his accent still noticeable after so many years in Canada. It was December 2017 and we were talking over coffee at a busy Tim Hortons in Maple Ridge.

The family picked out a small plot of land, made a deal and began building. First, the Paulokangases put up a rough shack and, of course, a sauna. They started a routine: almost every weekend, the whole family—Ray, his wife Seija, his young daughter Anita, his dad Niilo, and his mom Annikki, as well as his four younger siblings—would leave Vancouver and head for the cabin after work on Friday. They packed up two cars with roof racks for mattresses and lumber for Niilo's many projects. And then on Sunday around noon, they left again for the coast. "My dad never stopped working. He had a dream and wanted to make it happen," Ray said to me, wiping his pale blue eyes and shaking his head.

It didn't take long for Niilo and Ray to replace the shack with a sturdy five-hundred-square-foot cabin made out of logs they bought from a local rancher. "When we were building the cabin with my dad originally, we didn't have electricity. We stacked the

PRESSY LAKE–70 MILE HOUSE–GREEN LAKE

Ray Paulokangas lost the cottage that his father had spent thirty years building and perfecting.

logs, we would hand drill halfway down the log. We would fit in these six- or eight-inch spikes, with a strip of insulation between." Eventually they added a kitchen and a garage to the original structure. Ray said, "My dad worked thirty years without even going fishing on the lake. When the buildings were done, he was building rock retaining walls. He was one of those guys."

Four generations of Ray's family enjoyed the place. In the summer, they swam, fished and went horseback riding at the nearby ranches. In the winter, they skied, skated and played hockey. They took up ice fishing and used a smoker to preserve their catch. A few times they even spent Christmas at the cabin, when the country roads were passable.

"We used to get a lot of relatives from Finland and we always took them up," Ray recalled. "Those were probably the best times. We'd sit outside by the fire and have a few brewskies and go fishing. We would heat up the wood-burning sauna, get it up to about 80° C. And then we would go in the hot sauna, take as much heat as we could and jump in the lake. And go back. That's our tradition in Finland. So many good memories."

Ten years ago, after Ray retired from his electronics business, he and his wife bought a house and a nine-hole golf course on a forty-acre property at Tin Cup Lake, about twenty minutes away from the cabin. When Ray's wife died a couple of years ago, he stayed on at the property and carried on with the business. His kids came regularly in the summer to visit him and Pressy Lake. But all of it came under threat in the summer of 2017. On July 15, both the home at Tin Cup Lake and the cabin at Pressy Lake were placed under evacuation alert. Ray packed his clothes into big plastic tubs, took a photo album and some family pictures that were on the walls, and left to stay with his daughter on the coast. The alert was lifted on July 25 and Ray came back. But the all-clear was brief. On July 29 at 11:00 at night, an evacuation order was issued for areas south and east of Green Lake as well as 70 Mile House. The RCMP knocked on Ray's delft-blue door at 12:30 at night. "You've got fifteen minutes to get out," one officer said.

Ray hadn't unpacked much so it was easy for him to load up the truck again. For safety, he moved his machinery—the quads and the lawn cutters he had for mowing the golf course—to the middle of a field, away from the buildings. He summoned his dog but couldn't find his cat and had to go without it. Although he had been coming to Pressy Lake for nearly fifty years, he had never been ordered to leave because of a fire: "It felt like being evacuated from a war zone."

Half an hour after he received the order to leave, he stopped in at the 70 Mile General Store, a welcome beacon on that dark and frightening night. Thick, oppressive smoke was everywhere. While gassing up, he talked to his fellow refugees and learned that an evacuation centre had been set up in the curling rink in 100 Mile. He decided to drive there and register. When he arrived he found that cots were made up, but nobody was using them and he didn't feel like going to bed in that echoey cavernous space. One of the volunteers gave him a blanket and he slept in his truck for two or three hours. Then he headed to his daughter's place in Langley.

The traffic was bumper-to-bumper along Highway 24, a slow, steady stream of people towing their belongings. Their trailers were loaded with boats, quads, snowmobiles and animals. Sometimes, they had to pull over to feed their horses. It took Ray seven hours to get to his daughter's home, a trip that normally would take him less than five.

After a week went by, Ray, who was worrying about his cat, phoned the South Green Lake Volunteer Fire Department to see if one of their members could swing by and look for it. Ray explained where his keys were and a volunteer agreed to go over. "They went in the house," Ray recounted, "and phoned me back. 'Yeah,' they said, 'your cat is here because there are three or four dead mice on the floor.'" The volunteers put out a bag of food and a bucket of water, making sure Ray's cat was looked after.

"My place was protected very heavily by the local fire department. They went every second day to spray the place. On my property near Tin Cup Lake, there are three ponds that have water. They pumped them dry. They were very concerned. So was I. Every time I phoned to ask what was going on, the first thing they said was, 'Your place is okay. Don't worry.' The fire came very close, behind my property." Although he was well informed about his place at Tin Cup Lake, he heard very little about what was going on at Pressy Lake. Most of the residents had left, so there was no one he could phone for an on-the-ground report.

MANY YEARS AFTER RAY'S FAMILY started building the cottage on Pressy Lake, Alicia and Tim Polanski fell in love with a cabin they saw there while visiting their cousins, Wayne and Unni Lorentz. "My husband and I were out fishing," Alicia told me in a phone conversation. "I looked at him and asked him, 'What are you thinking?' He said, 'I was just thinking if I had a cabin, that would be the one I'd have.' And he gestured at a low brown cottage on a green bank behind a few trees. I went, 'So did I!'" The place they were looking at was somewhat neglected but "its bones were good," Alicia said. It turned out the owners were an elderly couple who no longer came up. For four years, the Polankis courted them and finally were able to arrange the sale in 2015.

Hoping to retire to the cabin, they spent a couple of years lovingly restoring it. Alicia said, "It had a real cute layout. Off the master bedroom, French doors led out to an eight-foot covered deck and then to a huge open deck. Doors from the dining room led to that same deck. I had huge pots of flowers on the deck. Everybody kayaking by would go, 'We love your flowers.' My husband had found beams on the property that were from old telephone poles. I remember he was coming across the lawn with this big smile, saying, 'Look what I found. These will be perfect.' He took out walls between rooms so the living room and dining room were all open. Then he took the poles and made big wooden doorways out of them. When you looked at it, it was just beautiful."

Originally, Alicia and Tim bought the cabin thinking they would use it mainly in the summer. But they discovered, as Alicia said, "Winter was our most favourite time. You had this beautiful little place to go to. You could go out in the cold and come back and be snug as bugs. My husband and I were out for a walk one day and my husband said, 'Stop.' So I stopped. He goes, 'Listen.' The sound you could hear was the snowflakes falling on your clothes. That's how quiet it is—the peace of it."

Alicia remembered the Canada Day weekend in 2017 with special fondness: "Our granddaughter was with us and we were sitting on the dock having a toast to it being the first time that everything was done. The last baseboard was in. The decks were done. The windows were all clean. We'd built huge rock walls. The gardens were gorgeous." Like Ray, Alicia and Tim had a place they'd lovingly made their own and were looking forward to enjoying for many years to come.

After the July long weekend, Alicia and Tim left their cottage to take their granddaughter to summer camp. They dropped her off but were unable to get back. The cabin was not under direct threat, but for some time, the roads to it were closed due to fires elsewhere. When the highways opened, the smoke was so bad that Alicia, who suffers from asthma, thought it best to stay home in Vernon. So they were not actually at Pressy Lake when the evacuation order was issued, although they certainly knew about it.

PRESSY LAKE–70 MILE HOUSE–GREEN LAKE

ON THE MORNING OF JULY 29, Lorne Smith and Cheryl Merriman were at Pressy Lake when they learned about the evacuation. Lorne said to me in a phone conversation, "We talked to the CO [conservation officer] and the police officer there and they said there was probably going to be an order. We were already on alert. So we left and went up to stay with some friends at Sheridan Lake."

Lorne and Cheryl had bought their cottage in the summer of 2016. After they retired in 2017, they left the Lower Mainland and moved up to the Cariboo. They had the benefit of their new home for just a month and a half when they had to go.

"Were you surprised when the evacuation order was issued?" I asked Lorne.

"I was shocked because the fire was so far away. It was a good twenty kilometres away. The fire didn't come through 'til the twelfth. There was almost two weeks before the fire got there."

"Did you think sprinklers would be put in?"

Lorne chuckled wryly. "That's a story in itself." Lorne and Cheryl were still at their cottage on July 28, when a group of structural protection experts came in. "The chief of the Comox Fire Department was leading them. When they saw us outside in our front yard, they stopped and introduced themselves. There were four or five people and they told us they were setting up sprinklers and pumps. They were looking for access to the lake so they could set them all up. They were in uniform and they had their trucks and equipment. I thought, 'This is great. If the fire gets here, they're going to have sprinklers on the roof.'" The next day, when Lorne and Cheryl drove north to Sheridan Lake, they trusted that in their absence the structural protection crew would keep their house safe.

ON AUGUST 12, ALICIA POLANSKI was still at her home in Vernon. She couldn't sleep for worrying about the fate of their cabin. She remembered, "I started reading about this wind event. Every time there was a wind event the fire went berserk because their guards were useless as tits on a bull. The wind event happened and at three in the morning the map changed. I wake my husband up and I say, 'You've got to look at the map. It's the devil; it's got huge horns.' My husband jumps out of bed and he's looking at the map with me. 'It's at the lake,' he says. I'm punching 'Pressy Lake' into my iPad. Trying to find out information. Out of the blue, there's the fire chief from Leduc in Alberta saying that at a little place called Pressy Lake, they had to pull the firefighters out and that structures were burnt. I ran with this thing 'cause I was going to show it to my husband and then the post was gone. Some of his [the chief's] people had been sent to the Pressy Lake area. What he had posted was that his people had to be pulled out and nothing was done."

A few days later, when Alicia got the call confirming that the fire had struck her property, all she could think about was how happy her husband had been to find those telephone poles, how he'd beamed walking toward her across the lawn to tell her the news. As she related this to me, she was overcome. I could hear her crying softly over the phone. "I had saved stuff from my grandparents, stuff from dad's cabin. My brother died. I had stuff from *his* cabin. They all went to the perfect home. I don't know what it is about cabins, but it's where your heart is." I understood what Alicia meant. I thought about our cabin. It had its collection too—watercolours of the "Little Island," kerosene lamps from Jane's family farm, a hand-carved trivet, a door frame where generations of children had wriggled themselves tall in order to be measured, their heights carefully recorded. These things all came with stories, roused memories of friends. I did not know what to say; I couldn't think of anything that would make her feel better.

Alicia said, "We had done all the fire safety stuff: raked pine cones away, trimmed and pruned. The wood wasn't stacked at the house. We had a steel roof. We had two hundred feet of fire hose, a generator and a water pump that burnt to the ground in our shed."

"Did you think your place would be protected?" I asked.

"Absolutely. They even wrote it in their reports. The TNRD would put the reports out, but they would come from all the people involved in the wildfires. You would see in the reports where it would say structural protection was going on to Green Lake, Pressy Lake, whatever. Now how the Sam Hell did Young Lake lose a freaking woodshed and all their houses were saved?" she asked me. "Very few houses on the lake, all back from the water.

Am I happy they were all saved? Absolutely. Am I royally pissed that we didn't get the same treatment? What the hell was wrong with us?" If anything, Alicia told me, the cottages at Pressy Lake should have been easier to save than the ones at Young Lake. They were closer to the shore and to the road. Neither access nor water was a problem.

While Alicia was in Vernon worrying about wind events, Lorne Smith had been waiting it out at Sheridan Lake. "I got called on August 18," he said, "and we were told our place had burned down. They told us it would be a couple of weeks before we could get in and see what had happened. It wasn't until September 16, a month later. There was nothing left." Both his cottage and two-car garage were totally demolished. "But we have insurance and we are going to rebuild. The insurance has assessed our house through pictures."

Ray Paulokangas came up in October with his son. His kitty had survived alone in the house at Tin Cup Lake for over two months. It had gained weight catching mice and chowing down on the kibble the firefighters provided. The machinery he left in the field was also unharmed, although it was now sprayed red with retardant. Planes had dropped fire retardant on one corner of the property where Ray had a barn. "When I got back," he told me, "I asked myself, 'Why is it so red out there?' I thought it must have been burnt, but then I realized it was the fire retardant."

Pressy Lake was another story. Just driving in was disheartening—kilometre after kilometre of blackened tree trunks. The fire was freakishly selective. It completely destroyed Ray's cabin, but the neighbours on either side of him were unscathed. Ray said, "I looked at the rubble there and all I could find were kitchen cups and spoons, a couple of tools, shovels with the handles burnt off. Then I found this one thing that we had on the bookshelf, an old man sitting on a rocking chair. I don't know where it came from. But I picked it up." Even Ray's aluminum boat was destroyed. It was sitting on a dock that projected out into the lake but the fire was so hot, the metal melted. "Everything was gone in two or three hours," Ray said.

Alicia was left with a canoe and a large carved cedar bear. She said the bear was about five feet tall, carved out of wood and standing erect as bears sometimes do when they are trying to hear, see or smell better. It is hard to understand why it escaped the general destruction (perhaps because it didn't have any branches to catch fire). All the trees around it were burned to the ground. She had two red plastic Adirondack chairs: "We had a place to sit, a boat, and a bear that should have been kindling." She also found some Christmas ornaments: "When the workers came to cart away the rubble, they lifted up the metal roofing that had caved in, and there against the cement was this little supply of Christmas things."

What went wrong? What happened to the structural protection that both Alicia and Lorne expected? When the residents of Pressy Lake tried to find out by asking BC Wildfire officials, they were told to file a Freedom of Information request. Heather Pederson undertook the task of submitting the request. She has a cabin at Pressy, and although it survived she still wanted to find out what happened. She is a law enforcement officer in Coquitlam; Alicia describes her as "my Nancy Drew." Heather filed two applications, one for BC Wildfire Service records about structural protection units at Pressy Lake and another for wildfire investigation reports from the Office of the Fire Commissioner. But she ran up against bureaucratic roadblocks and asked Jackie Tegart, the MLA for Fraser-Nicola, to help.

On the morning of September 21, an angry Tegart stood up in the BC legislature. She said, "Ms. Pederson has been told to submit an FOI request if she wished to know what measures were taken to protect her property, and she did just that. Since then, she's been contacted twice to withdraw her FOI request because it's 'difficult to obtain' and 'likely to be part of some sort of investigation.' These people have hit rock bottom. Their homes are in ruins. Livelihoods are gone. Can the Minister of Citizens' Services commit today to ending the intimidation and ensuring the residents of Pressy Lake get the documents they need to understand exactly what happened?"[55]

Jennie Sims, the Minister of Citizens' Services, responded, "There isn't one person in this House who does not empathize and is not feeling the pain that British Columbians are feeling as a result of these fires. We want to get to solutions, and we can get to them when we sit and talk. So come to my office, or I'll come to yours, and let's get this sorted out." Sims vowed to expedite information for those affected by the wildfires.

On October 31, Heather received five hundred pages of reports with a pledge of more to come. They are now posted on the internet, which is where I found them. They include a logbook written by Glen Burgess, who was the incident commander of the Elephant Hill fire at the time, handwritten notes from ground crews, weather reports, photographs, diagrams and structural protection summaries. I also found something called "A Response Package," which contains 165 pages of emails, daily situation reports, triage forms and photographs. Many documents are redacted, so key information is missing, but by studying them carefully it is still possible to glean something of the tragedy that lies buried here.

The fire chief from Comox who had introduced himself to Lorne Smith was Gord Schreiner. At 7:00 a.m. on July 28, Schreiner and a protection unit known as First Call left Clearwater, where they had been working, and went to Pressy Lake.[56] They were dispatched because of concerns that the Jim Lake fire, about four kilometres northwest of Pressy Lake, might pose a risk. The documents show that for some time anyway, the needs of Pressy Lake residents were taken seriously. I found detailed assessments for eighteen different properties. They were all labelled "Triage #1" and the documents indicated what each place needed to be safe: one property required two rooftop sprinklers whereas another needed four, and so on. The structural specialists made diagrams that noted the position of primary buildings as well as sheds and boathouses. They commented on special hazards such as propane tanks. They also took photographs. The copies I saw of them were grainy, but nevertheless it struck me how well tended most of the cottages were. The grass was cut, the lower branches of trees were trimmed away. Porches didn't sag, often roofs were metal.[57] Then I learned that on July 29, Schreiner and First Call were "released" from Pressy Lake. The nearby Jim Lake fire had grown to forty hectares but was "contained" and Pressy Lake was considered to be out of danger. First Call went to Clinton and Schreiner travelled home to Comox. Once home, he sent an email to Rick Owens, a fire service advisor, confirming that he was released from his duties at Pressy Lake.[58]

But the situation was fluid and though the danger to Pressy from the north was over, the Elephant Hill fire to the south continued to rampage. A few days later, another crew went back to Pressy Lake because of that threat. Among the FOI documents that Heather received was an unsigned, handwritten note labelled "Pressy Lake Assessment," dated August 7. It summarized the needs of the residents; a total of 71 primary structures required 8 pumps, 156 sprinklers, and thousands of feet of hose.[59] Other documents indicate that sprinklers were installed at Clinton, the West Fraser Mill at Chasm, 70 Mile House, Highway 99, the Highway 1 corridor, Hihium Lake and Young Lake. But Pressy Lake was not mentioned again.

When the Elephant Hill fire blazed into Pressy Lake on the night of August 12, firefighters from Alberta were dispatched to the scene, but they found a raging inferno so violent, they withdrew. At 8:30 on the morning of August 13, Glen Burgess wrote in his log, "Have confirmed structure losses—nothing being said publicly yet, internal intel only—losses at Pressy Lk, fifty knot winds."[60] Structural protection specialists who are identified as SPS 107/114 write in an incident report dated August 15, 2017, "Rank 6 fire behaviour was observed and numerous properties were destroyed on the initial night of the fire taking over the area."[61] On the BC Wildfire Service's scale of severity, a rank 6 fire is the highest. In its words, it is "a blow-up or conflagration; extreme and aggressive fire behaviour."[62] Pressy Lake was left to burn and thirty-three structures were destroyed.

Despite the FOI release, many questions are left unanswered. Why did no one act on the August 7 report? At that time, firefighters would have been able to work in the area without risking their own lives. Was the Wildfire Service short of equipment? Of firefighters? Did other lakes have higher priority for some reason? As Lorne Smith said, "They protected Green Lake like it was gold."

The government initially resisted releasing any documents about Pressy Lake. Understandably, people began to suspect a cover-up. There were persistent rumours of a back burn set between Young Lake and Pressy Lake that went awry. Lorne said,

"I talked to the BC Wildfire Service via email and I asked them directly, 'Did you put a back burn in?' They said, 'We can't comment on the back burn.'" Lorne believes the burn may have been set on the Lost Valley Road. However, unlike the back burn south of Clinton that went amiss, which is well documented by photos of helicopters dropping incendiaries, there is no such evidence of this one. The FOI documents mention burns planned for many areas, including Young Lake, Clinton, the West Fraser Mill at Chasm, 70 Mile House, Pavilion and Skeetchestin, but not Pressy Lake.[63] Yet the troubling hearsay endures, leaving us with more unsettling questions. From the pages of material that the FOI requests have so far unearthed, we know *when* structural protection was pulled from Pressy Lake, but we're not much wiser about *why*.

Lorne said, "I still want to know what happened. I want the truth so we can move on. If they had said, 'We screwed up. This is what happened,' you know what, that's fine. At least we'd know. I could accept that."

On September 16, the day Alicia returned to Pressy Lake to see the damage with her own eyes, she was cleaning up and dragging some branches, limping badly due to an ankle injury she had suffered earlier in the year. A firefighter called Joe, who was helping with mop-up, came by and said, "We're going to come in and help you." Alicia said, "Oh no, we're good." Joe went away but he returned with a few of his colleagues, who began cutting down danger trees, those likely to fall over due to fire damage.

After they were done, Alicia was standing with Joe looking out at the lake. As she was telling me about this, I could hear her voice breaking at the memory: "The lake looks like the Black Hole of Calcutta," she recalled, "and he puts his arm around me and goes, 'Don't despair. The ash makes good fertilizer. You'll be surprised.' But I can see tears in his eyes and he goes, 'We tried so hard to make sure that this didn't happen.'"

When I first met Ray Paulokangas, he said, "We intend to build a new place." But zoning bylaws have changed since his father's time. New rules about setbacks from the lake for cabins and septic fields make it challenging if not impossible to build on a small lot like his. Even though it was a legally constituted property, reconstruction looks to be much more difficult than he expected. In the summer of 2018, Ray announced on Facebook that he had decided to go back to Finland, where he was born. "The last two-plus years have taken a toll in my life and I feel I should live the remaining years of my life in my homeland where my roots are." After living in Canada for fifty-three years, this meant leaving his children, his grandchildren, his brothers, his sister and their children and grandchildren. Grief about his cabin was part of the reason for his choice. I was so sad when I heard about it. I told him I thought Canada had failed him.

The folks at Pressy Lake had believed they would be protected. Their broken trust will not be easy to restore. And I wonder how stories like this will influence other residents at other lakes in future fires.

CHAPTER 21

WILL YOU PICK UP BEAR?

WHEN 100 MILE HOUSE WAS evacuated on July 9, Miguel Vieira's family, who live there, needed a place to stay. They went to be with Miguel, who lives at 70 Mile House. His wife, Krista, said to me, laughing, that, as he is Portuguese, "he has quite a bit of family." Nine of Miguel's relatives came, and then Krista's parents, whose place is on the south side of Green Lake, also arrived. "Our driveway," she recalled, "turned into an RV park."

As this one example reminds us, many British Columbians who had to leave their homes did not register at evacuation centres but went to stay with relatives and friends. Official reports peg the number of displaced people at sixty-two thousand,[64] but not all refugees were counted and the total tally is likely significantly higher.

Krista and Miguel Vieira are the owners of the 70 Mile General Store, which is where Gordon and I met them for several hours in February 2018 to talk about the summer of 2017. "It wasn't our first fire," Krista said. For a few days back in 2009, folks a couple of kilometres to the north were pushed out of their homes due to a wildfire. "Everybody came to the store," Krista said. "They camped out here in their trailers and vehicles and we made breakfast by donation." During that fire, she also looked after the children of single dads who were volunteer firefighters. "We had six kids at our house, sleeping everywhere."

On January 2, 2010, a fire began in the ceiling of the store. By 3:00 in the afternoon, the building was engulfed in flames, but fortunately no one was hurt. The Vieiras rebuilt even though it took a year to do it. Krista had grown up in the store—she was three when her parents bought the business in 1987 and wasn't going to give up on it lightly.

The evacuation of 100 Mile House caused many ripple effects. While most of the residents left the town, not all of them moved from the area. Many went to friends and relatives nearby, and still needed food and other supplies. Because 100 Mile was out of commission and several roads were closed, suddenly a crazy-quilt situation developed. A few stores were besieged whereas others, fully stocked, were devoid of customers. The state of affairs caught the attention of Ann Hui at *The Globe and Mail*, who reported on July 13, "Since receiving its last regular shipment last week, the Lone Butte store has been mostly sold out of milk, bread, eggs and fresh produce. The store was able to get some supplies on Tuesday after a truck destined for Lac la Hache, about an hour north, was turned away from reaching that town." Meanwhile, John Sperling, the owner of the Safeway at 99 Mile House, just south of 100 Mile, was fully stocked but potential customers could not get to him. The evacuation order meant not many residents were left in town and the blockades were preventing anyone else from entering it to purchase food or other necessities.[65]

In anticipation of being besieged, the Vieiras put in a huge order with a variety of suppliers. On July 11, their garage was filled to the brim with groceries. "It was just insane," Krista remembered. For a while, the store did twice as much business in a day as it normally does in a week. Miguel's family helped by stocking and restocking the shelves, but it was hard to keep up. Normal supply chains were disrupted. Milk was difficult to get and chips, that staple road food, also were elusive because they normally came from 100 Mile. The Vieiras couldn't get fresh hot dog buns from the bakery that usually provided them. And the buns they bought as a substitute were not only less desirable (frozen), but much more expensive. "They *cost* more than what we were selling them for previously. But we couldn't up it. We just took the cut." Krista said. People have long memories in the Cariboo. During the fire that hit 70 Mile in 2009, one local motel upped its prices to over $100 a night. "We saw how much it affected the business—still. I was like, 'We're not upping prices,'" Krista said.

On July 15, Clinton and South Green Lake were placed on alert; Williams Lake and an area east of Clinton were evacuated. The fire activity, especially to the south, made the Vieiras' family nervous. Except for Miguel's sister and brother-in-law, they all left to get farther away from the fires. Krista worried about her two girls, Makayla and Hannah, who were nine and seven at the time. When a friend in Port Moody, who also had two girls, offered to take them, Krista happily accepted, relieved to get them out of the fire zone.

But many of the properties to the south of 70 Mile had their orders downgraded to alerts on July 20, and two days later the residents of 100 Mile House were allowed back. Krista relaxed a little. Things seemed to be returning to normal. Miguel's family and Krista's parents went back to their own places and the girls came home from Port Moody. Tides of people were washing in and out along the roads of the Cariboo. When the fire blew up, they ebbed out and when it died down, they washed in.

For two weeks, Krista and Miguel were working so hard they had not left their property. They decided to take a short holiday and drove over to Green Lake to see Krista's parents. On July 29, the family went to 100 Mile House to attend the funeral of Miguel's uncle; Krista and Miguel fully expected this would be a short interruption of their holiday. They anticipated going back afterward and enjoying a few more days of summer at the lake.

Events intervened, however. While in 100 Mile, Krista said, "I got a heads-up from a few different people in a few different ways. One was our TNRD rep, Sally Watson, who was like, 'Looking like we're gonna get evacuated. It would be awesome if you stayed open.' I was like, 'Yep, not a problem.'" Krista realized that people were going to need gas and food. She figured they wouldn't be stocking up on staples, but they'd want pop and chips. She said, "They're all driving in the middle of the night. And what do you do when you drive? You snack." As they expected to be unusually busy, Krista and Miguel left their kids with his parents in 100 Mile. "Don't worry, we'll come back and get you in a little bit," Krista told her girls. She did not for a moment imagine that 70 Mile would be placed under evacuation order and that the separation would last for much longer than "a little bit."

Miguel is a member of the 70 Mile Fire Department, and he was called to take a shift at the sawmill in Chasm, ten kilometres to the south. His group of volunteers and a structural protection crew had been dispatched to set up bladders, pumps and sprinklers. That left Krista alone to manage the store. Luckily she was able to get a few people in to help. Krista recalled the night of July 29 vividly: "This was when shit hit the fan. That was the Big Crazy. This was when Green Lake and Pressy Lake were evacuated. All those places—everywhere, everywhere, everywhere—from Clinton north. And they were all told: 'Highway 24.'" From 4:30 p.m. on, the TNRD issued four orders and one alert. It was like a rolling blackout. The last notice was issued at 10:00 p.m. That's when Krista learned that 70 Mile was under order too.

For some people, The Knock came at 1:00 a.m. An RCMP officer standing outside. A bleery-eyed resident quickly tying on a robe. "You gotta pack. You gotta go now." No time for niceties, for chats, for the delicate exchange of information, for explanations. "You gotta go." As people drove north in the darkness, they encountered no traffic coming south. Only their own headlights cast a wavering beam into the smoke. Ash rained down; behind them, the sky glowed an ominous red. The smell of wood burning was inescapable.

"We would get people coming into the store: 'Oh my God! Oh my God!' It was the pure relief that the store was still open," Krista recalled. I understood completely. You could choose a chocolate bar, get a drink, make a purchase. Something ordinary, something you do every day, rather than running for your life and worrying about how much destruction you would encounter when you got back. What would it be? Scorched earth? A cabin reduced to rubble? Instead: *Hey, the lights are on. A familiar face behind the counter.* Krista said, "I don't know how many people I consoled. I don't know how many hugs I gave."

In the midst of the exodus on July 29, she had a realization that knocked her breath away. Her daughter Hannah's stuffed bear, called Bear, and her blankie were in their trailer at her parents'

Krista Vieira with her daughter Hannah. Beloved Bear was left behind the lines—and rescued!

place at Green Lake—behind the lines and inaccessible. "So our daughter doesn't go anywhere without Bear," Miguel interjected, looking at me solemnly and then pausing to let the full effect of this situation sink in. As a parent who had years ago undertaken several retrievals of stuffed animals, I comprehended immediately.

Krista is nothing if not resourceful. She discovered that Carolyn, a friend of hers, was headed out to her daughter's place at Green Lake to retrieve some things. A few people were apparently still being allowed to assist with the removal of property. Carolyn had worked at the store when Krista was three. "That's how long I've known her," said Krista. "So I asked, 'I know it's silly but will you pick up Bear for me?' 'Oh honey, 100 percent,' she said." Krista explained where Bear was and at about 4:30 in the morning, Carolyn arrived with Bear and Blankie in hand. Krista wept with relief. "At this point I was a mess. I had consoled everybody else, but I was toast." Half an hour later, she finally went to bed.

This wasn't the end of Bear's saga, however. Between Bear and Hannah lay not one but two barricades—one checkpoint at the crossroads of Highway 97 and North Bonaparte Road and the other at the junction of Highway 97 and Highway 24. Krista did not have a pass. If she went through a checkpoint without one, she would not be allowed back in. But she found out that some of the RCMP officers who were working in her area were actually staying in 100 Mile. So the next day, she asked one of the officers, a woman, as it happened, "Would you please drop this off?" Without hesitation the officer said, "Yes, I can do that." Krista gave her the bear and Blankie as well as an address, and that evening she delivered both to Hannah.

"Must have been the funniest delivery ever for an RCMP officer," I said. "She was a little saving grace," Krista replied.

That day, July 30, was memorable for another reason: Krista saw the Canadian Armed Forces roll into 70 Mile. It was about 7:30 in the morning and she'd had only a couple hours sleep. She was struggling to rouse herself to alertness when the tanks arrived. "How did you feel, when you saw them?" I asked. "It was just like shit just got real. Like it just got real. I thought it was kind of cool. People said to me, 'How can you say that it's cool?' It was like, 'In all honesty, my life is pretty rough right now. Haven't seen my kids for a while, haven't whatever.' So you want to know what? I took five seconds and thought to myself, 'Wow. That was cool.' Because in that moment there wasn't a lot cool. No, there wasn't a lot positive."

Miguel, who had been working all night on structural protection at the Chasm sawmill, had come off his shift and was driving home. He'd been up for twenty-four hours. "I was in zombie mode," he said. He'd just reached 70 Mile when he saw the line of army trucks rumbling south. The Canadian army had left Williams Lake to help the RCMP with evacuations and staffing at the checkpoints farther south. "There were close to fifty vehicles that came down the road. I sat there for five minutes watching," Miguel said. Probably the most striking were the armoured personnel carriers, some complete with a gun turret. They looked like tanks, but instead of treads they had four wheels on each side. Later, Krista took a picture of a soldier in one of these vehicles, pulled up to the gas bar in 70 Mile, waiting for a fill up—a most unusual sight for Canadian eyes. Krista became friends with the fellows she called "my army boys." They put up tents across the highway from the store and Krista let them use her washroom and shower. "We got close," she said. "One guy had a baby and showed us the pictures. Then he came back for another stint and he showed us updated pictures."

The store remained open while 70 Mile was under evacuation. Not only did it continue to supply groceries to the fifty or so residents who had remained behind, but it was a hub of information. Good information was at a premium, so that was critically important. Ray Paulokangas, who was evacuated from his home at Tin Cup Lake on the night of July 29, made a point of stopping at the store after leaving his house. By talking to other evacuees as he was gassing up, he found out that the only way out was north and that he was supposed to register at the evacuation centre in 100 Mile. At one point, an RCMP officer with whom Krista was friendly warned her that the police were going to be stricter about making sure that residents who were still staying in 70 Mile kept to their own property, The officer said, "While North and South Green Lake are on lockdown, 70 Mile is on do-whatever-you-want." The officer made clear that things would become a little tighter and the rules would be enforced. When she was talking to people about their food orders, Krista passed on the message. She kept a Facebook page on which she posted the latest news, from official government and other sources. She also gave people her cell phone number if they wanted updates. She told them, "I don't want to talk, but send me a text and I'll give you an update. There were a lot of people at South Green that had no information. It was so hard that way."

The fire never got closer than fifteen kilometres away from 70 Mile, but when the wind blew, Krista always worried that it would come through. And the Elephant Hill fire *had* been known to travel fifteen kilometres in one day. "We tried to make the best of it," Krista said. She worked at keeping up her morale and that of others. Although her "army boys" had enough rations to survive, she sometimes gave them a home-cooked meal or brought over coffee or peaches or cherries. "Alicia was one of our staff that was here the whole time," Krista recalled. "About the end of August she said, 'I really see how much the store influenced the whole community.' Maybe that's because I've grown up here in a small town and I know what it's like. It could have been a whole different feel around here. We could have just stirred this up to a little festering pot of negativity. But instead we delivered peaches."

At last, 70 Mile got the all-clear on August 15. The Dusty Rose Pub opened for business that day; the next day, Makayla and Hannah came home after eighteen days away. This time, the village almost returned to normal; full recovery, however, would take much longer.

The Vieiras had a few astonishingly madcap-busy days, but overall, business was down during their fire summer. I can report, however, that when we visited in February 2018, Bear was still thriving; I noticed Hannah solicitously washing him in the sink. And in a follow-up phone call, Krista told me that in August of that year, business was very lively due to the mushroom pickers who came in to take advantage of one of the fire's parting gifts—a huge crop of morels.

CHAPTER 22

I COULD WRAP A WET TOWEL AROUND MY HEAD, GO OUT INTO THE LAKE

"WE BUILT A RESORT ON Green Lake, a unique spot. I didn't want to let Mother Nature take it without a fight. Everything I've wanted to do in my whole life was all here," Brad Potter told me. He elected to stay on his property during the fires of 2017. "I felt I knew what I was doing," he said. "I'm an ex-fire chief and my wife and I founded the Interlakes Volunteer Fire Department. I was either president or fire chief for nine years. I've fought wildfire and lots of structural fires. I had two fire pumps, tons of hoses."

Brad and his wife, Gail, own a home-cum-lodge known as The Wind and the Pillows. They had their seventy-six-hundred-square-foot post-and-beam structure built in 2011 by Pioneer Log Homes of BC, a company featured on HGTV's show *Timber Kings*. The house is on a peninsula on the southwest shore of Green Lake. Although you might not easily see it, the area is steeped in history. For centuries, Green Lake was a gathering place, the site of an annual meeting "where trade, political discussions, kinship ties, sports, spiritual, ceremonial and communal activities were shared and celebrated by Secwepemc from near and far."[66]

Brad has been a realtor in the South Cariboo for twenty-five years. He and Gail are also enthusiastic musicians. Playing music together in high school drew them together and they still have a band, The Classmates, a popular choice for local events. Brad is talkative and gregarious. We spoke on the phone for an hour, and Gordon and I stayed at the lodge for a couple of nights in February 2018 to get more of a sense of what he had defended.

Brad told me that when an alert was issued for Green Lake on July 15, the South Green Lake Volunteer Fire Department came and said, "Get the hell out," even though an alert doesn't require you to leave, only to get ready for a possible departure. Brad and Gail gathered up their valuables and artwork and packed them into their boat, which was sitting on a trailer. The next morning, Gail said, "Let's go, Brad." He said, "I'm sorry, I can't leave." He handed her the keys. Gail hates driving with a trailer, but she didn't argue with him; she just gave him a hug and left for Kamloops.

A week later, the two decided the fire was quite far away and the chance it would come as far as Green Lake was remote. *It's not going to happen*, Brad thought. The Elephant Hill fire was estimated at fifty-eight thousand hectares, but on July 22 the *Ashcroft Journal* reported that for the last forty-eight hours, the fire hadn't grown much and was staying within its perimeter.[67] Gail returned. However, on July 29 an evacuation order was finally issued for South Green Lake. The Elephant Hill fire had grown to seventy thousand hectares and was breaching its containment lines north of the Bonaparte River. Its meekness of a few days prior had quite vanished. At midnight, the Potters' phone rang. Brad said, "It was my daughter in Edmonton saying, 'Get the fuck out of there. You've been evacuated.'" Fifteen minutes later, a fire truck pulled in and gave the Potters the same news. Once again, they hooked up their truck to the boat, which was still packed with valuables. Brad drove the truck and Gail took the car to their friends' place on Horse Lake. Gail stayed there, but Brad went back home. "That was the start of my siege," he said. He knew that once he decided to stay behind, he could not leave the premises. "The word was if you were caught off your property it was automatic—two nights in the can and you weren't coming back. I know people who were caught and were put in the can."

Then Brad got to work preparing for the worst. He had "fire-smarted" his own property. But the parkland around his place was a problem, as it turned out. Green Lake Provincial Park consists of eleven distinct sites scattered around the fifty-seven kilometres of lakeshore. While some of the sites are designated for camping, the Boyd Bay site, which surrounds the Potters' place, is zoned as "natural environment." According to the regulations, that meant Brad was supposed to let natural processes proceed unhampered. He wasn't allowed to mow the grass or dispose of any dead trees. These, of course, were precisely the kinds of preparations that homeowners were being encouraged to undertake to prevent fires from spreading onto their properties. "We'd had a wet spring and everything grew quite lushly. And then there was no precipitation in this part of the province and everything died. I went, 'If there's a spark there, that's an instantaneous fire.'" Brad ignored the rules. He decided to cut down the hayfield next door—"a quarter of an acre of grass that was two-and-a-half-feet tall, all dried out." He had a mower with one rotary motor and a bag on it. He'd run it for forty feet, the bag would fill up, he'd empty it, and then he'd run it for another forty feet. It was hot, about 30°C, and the work was slow. But after a week, Brad had the hayfield cut back to lawn length, "I had two fire pumps and hoses and then I started sprinkling the crap out of everything."

"It was at noon one day," Brad said, "when a truck pulls into my property. It was a fire truck pulling a twenty-four-foot trailer with roll-up doors down the sides. You roll up the doors, there's tray after tray, row after row of sprinkler heads and little hoses." A crew had come to evaluate the main buildings on Green Lake for structural protection. "Then they saw what I had set up and said, 'Christ you've got 98 percent of it, so we're going to do your house right now.' Five guys, twenty-five, thirty years old. They hop out and climb all over my buildings. They put sprinklers on the four peaks of the house, two on my shop, one on every cabin and every building, and hooked up to my fire hoses and the pumps. They supplied the sprinklers. Then they left to do the next-largest place. But before they got there, they got called down to Clinton. That was when the fire was attacking Clinton aggressively again. I had my house sprinklered a week and a half before anyone else up here. It was another crew that eventually did the rest of Green Lake."

Brad spent his days moving sprinklers. Eventually, the brown grass on his property turned green. Every six days or so, the wind would pick up, and with 30–50 kph gusts behind it, the fire would just take off. One such occasion was Friday, August 11. In his work log on August 10, Glen Burgess, the incident commander for the Elephant Hill fire, wrote, "Discussed the Wx change the next few days and potential spread to N, NE, E." ("Wx" is an old abbreviation for "weather," dating back to the time of telegraphs.) His August 11 entry at 14:10 reads: "Div. B, fire activity increased pulling crews to safety—fire moving northerly—East Branch, fire activity picking up."[68] That day, a printed Fire Behaviour Forecast from the BC Wildfire Service predicted winds of 30 kph in the region by late afternoon and warned about an intermittent crown fire with flame lengths of over nine metres.[69]

Brad Potter, who owns Wind and the Pillows at Green Lake, is decked out in firefighting gear to defend his home.

"It was just a wall of inferno five miles long," Brad said, "basically ran along the south side of Green Lake, a couple of kilometres away. That was the worst night. You don't know how far away that fire is. I'm surrounded by trees. I used to go out at dusk on my ATV. I'd go out to a gravel pit down the way and get on a high hill so I could at least see a distance and figure out where things were. So I was awake all night and the pumps stayed on all night. I went through $500 worth of gas just to keep the pumps going. That night was just incredible. All the smoke that's being produced by this fire ends up looking like fire too, because of the reflection of the fire in the smoke. You could see the trees candling and the flames going way up into the air. Gusting winds, it's an inferno. Pretty scary."

On August 12, Glen wrote in his first log entry of the day, "concern with people not leaving, taking advice from so-called 'experts,' not us."[70] That situation, and the disaster that had just occurred hours before at Pressy Lake, may explain why, at 11:00 in the morning, Brad saw two pickups, carrying two conservation officers and one RCMP officer, pull into his place. "About ten people on Green Lake had stayed and the conservation officers and the RCMP officer had the job of convincing them to get out. Brad said, "These are three grown men who were scared shitless. You should have seen them. It scared me, they were so afraid. They said, 'Get the hell out of here. The fire has jumped all the guards that they have been building for the last two weeks. It's coming here, get out of here.' I said, 'No. I made my decision. I'm here.'"

But when the conservation officers and the policeman left, Brad didn't feel so good. "I sat down for half an hour. It was the only time I was scared." *What am I doing?* he asked himself. "When a fire comes at you like that, you can't fight it. But I did have a Plan B: I had a bottle of whiskey in my boat. I could wrap a wet towel around my head, go out in the lake and watch what the hell's going on. I went, 'No. I made my decision. I've been here for a friggin' month and I'm staying and getting back to work.'" If Brad had known about what had happened at Pressy Lake, the advice from the RCMP and conservation officers might have given him pause. However, the story about the arrival of an explosive rank 6 fire, which drove the firefighters away and burned so many homes and cabins at Pressy Lake, would not be confirmed publicly for another week.

After his visitors left, Brad did more clearing and watering. That evening, he put on his full retardant fire suit and pressurized mask and got on his ATV. He drove down Hutchison Road, where a big long finger of the fire was pointing straight toward him. "I just had to know what was happening," Brad explained. "Nobody was telling anybody anything and I couldn't see." It was getting dark, about 9:00 p.m., and as he drove south and approached the epicentre of the fire, more and more smoke enveloped the road. On either side, he could see flames here and there, stumps and snags ablaze. He got as far as Hutchison Lake, about eight kilometres to the southeast. "It got so smoky," Brad said, "I couldn't even breathe with my respirator on." Still, he concluded that the intensity of the fire was ebbing. "It was smouldering all over the place and if a wind came up, it would be there again." But for the moment, he felt the beast was slumbering and he could go back and get some sleep. He said, "I'm sixty-four now. I couldn't stay awake two nights in a row like I used to." Gail, at Horse Lake, stayed up all night, glued to Facebook so she could see if something came up. If it did, she would phone Brad and alert him.

On August 13, the BC Wildfire Service predicted cooler temperatures and even a few scattered showers. Brad settled back into his routine. In the morning, he told me, he'd see a little smoke, which by 11:00 would turn into a plume. "By 12:00, it looked like Hiroshima just south of us and then at 4:00, the water bombers would come in. You had two sets. You had the Quebec bombers, the big yellow ones. They fly in fours." Known as CL-415s or Superscoopers, the planes were specifically designed for aerial firefighting and perform well in the gusty conditions often found around forest fires. They can heft six thousand litres of water per drop.[71] "Then you had the Fire Boss planes, flying in sixes," Brad said. They are smaller, capable of dropping three thousand litres of retardant or water. They can fill a tank in fifteen seconds from over seventeen hundred bodies of water in BC.[72] "They would all use my house as the staging area," Brad said. "Every afternoon for a couple of weeks, as soon as they arrived, these planes circled my house four or five times just to get lined up properly. They're looking at the air currents and the waves and to get their orders about where they're bombing. And of course, I'm out there waving at them and they're dipping their wings at me. I swear they were trying to touch my house sometimes, they were getting so close. They'd go off and knock down this fire and the smoke would go away and the next day at 11:00 there's smoke again and by 12:00 it's a raging inferno and the fire bombers would come back and do it again."

It is remarkable how much water these planes can dispense. The Superscoopers can drop fifty-four thousand litres per hour when the water source is eleven kilometres away. In that time the Fire Bosses can dispense fifty-three thousand litres. If the full fleets of four and six planes are running for eight hours a day, that adds up to over 4 million litres of water. About the same volume of water goes over the Horseshoe Falls at Niagara in two seconds.[73]

While the fixed-wing aircraft were dropping their loads of water, sixteen Bell 212 helicopters were launching from the Flying U Ranch on the north side of Green Lake, fourteen kilometres away from Brad's place. "They're fifteen-passenger helicopters," Brad said. "They're all over the place. You also had the Sikorsky Skycranes—two of them. They were trying to fight the fire behind the main fire line, to put out all the hot spots. They'd hover over a pond, suck up all the water and frogs and drop it on a snag here and there. You've got sixteen helicopters going all the time, bombers flying around, bird dog planes." As I mentioned in chapter 11, bird dog planes provide a mobile, in-air traffic-control service.

"And then," Brad said, "you have the retardant bombers." One day, Gail, who was constantly scouring the internet for information, got a Facebook message from a friend, who was at the airport in Kamloops sitting next to a senior forestry official. She somehow discovered that bombers were coming to drop retardant on the Potters' house. Gail phoned Brad to let him know and sure enough, a few minutes later, four planes showed up: two propeller-driven and two jets. I thought this was an interesting use of social media, illustrating the speed at which it can disseminate news. And though Facebook is often rightly criticized for permitting the spread of inaccurate stories, there is nothing *inherently* inaccurate about it. Here you see how ordinary people can use it to access intelligence that previously was unavailable to them.

Brad recalled the planes' arrival vividly: "They're circling my house but way up high. My wife is phoning me: 'Get inside, you can't be outside.'" Gail was worried about the toxic effects, but Brad wanted to *see* what the planes were up to and he figured he could manage the risk. "I went outside. My wharf is a hundred feet long, so I went out to the end, parked my butt on the end of the wharf, got a case of beer and got all my cameras. I sat in the boat for forty minutes while these big friggin' vultures surrounded me. They kept circling me. It was kind of freaky. My property is on a peninsula. What they did was to pick the narrowest part on the peninsula. They finally came in and dropped three loads of retardant. They've got the bird dog plane and four friggin' bombers. And I'm going, 'I've got my tax dollars back for the last ten years.' You just think of it. It's a lot of money. I find out after the fact that every time they drop a load of retardant, it's $10,000. That was $30,000 they put across the narrow spot on the peninsula. It was crazy. Quite an airshow all summer long."

AT THE BEGINNING OF SEPTEMBER, Brad managed to sneak his wife home. Then on Tuesday, September 12, Brad's birthday, the official word came. He got the present of his life when he found out that, at 3:00 in the afternoon, the all-clear would sound. After six weeks of lockdown, the RCMP opened the checkpoints to let everyone back to South Green. "We made a big sign," Brad said. "My wife uses a golf cart to get around the property. We'd funkied it up because the name of the resort is 'The Wind and The Pillows.'" The Potters had altered the cart to make it worthy of Mr. Toad, a character in *The Wind in the Willows*, giving it big bug-eyed headlights, a roof with LED lights going around it, and a stereo system. When the Green Lake residents came in, it was a regular parade, a police car leading the way, driving very slowly, people honking their horns. The Potters went up to the road in their golf cart, cranked up the salsa music and waved as their friends and neighbours arrived. On the weekend, they had a big pig roast and potluck dinner. Bands played and 175 people came. "It was a hoot," Brad remembered. "And that was the end of it all, I suppose, although things were still burning up in the hills."

The Potters' house survived, but they didn't get off scot-free. Brad owns sixty-five hectares down on Hutchison Road that was totally wiped out by the fire—a quarter section sitting at the end of a lake, surrounded by Crown land. Brad told me the original Begbie Trail, built in 1843, went through it. I couldn't find a reference to a "Begbie Trail" of that date. In fact, I read that Mathew

Begbie only arrived in the colony of British Columbia in 1858 to take a position as the first chief justice. But Brad got the date right. I did find that a Hudson's Bay Company trader called Alexander Anderson used a trail between Green Lake and Horse Lake from 1843 to 1848.[74] This trail predates the Cariboo Wagon Road, which was built in 1862, and followed the route of the original Hudson's Bay Brigade Trails. Brad said that the remains of a way station—two log cabins and a log barn—were still on his property. They date back to the Cariboo Wagon Road days, when there were stopping points every ten or eleven kilometers so the coaches could change horses. "All that's gone," Brad said. "The fire was so intense that even the topsoil burnt."

Brad's struggle with the fire was basically a success story. However, even he had losses. We all did, really, when a small part of British Columbia's history went up in smoke.

SHERIDAN LAKE

CHAPTER 23

IT COME IN HERE WITH A VENGEANCE

BY AUGUST 17, THE ELEPHANT HILL fire had grown to 168,000 hectares.[75] Its progress north was relentless. Gordon and I were staying at our cabin on Sheridan Lake and the perimeter fire map showed a finger pointing northeast directly at us. The tip of that finger was only ten kilometres away. I wrote in my diary:

> Still very worried. Supposed to be big winds tomorrow. Mark texted Gordon they would be 80 kmh. I don't see that on the internet weather channels—Weather Network and AccuWeather. I have moments of fear—moments of okay. The fear is often prompted by others: Mark and his 80 kmh winds; Ann, who said there would be a storm this aft (there wasn't); also Larry's description of fires didn't help, flames shooting three hundred feet up, generating winds that send embers tumbling onto houses—scary!

We undertook preparations. I always felt better when I was doing something rather than simply waiting like a sitting duck. My diary noted, "We bought a fire hose and Gordon had fun wetting down the cabin and environs."

August 18. The skies were clear and the winds from the northwest. "Today we'll be painting the house with fire retardant," I wrote. "Still loads of downed trees to deal with. What are we going to do with them all?"

August 19. The perimeter map had not changed. A finger was still pointing straight at us and was still ten kilometres away.

> Predicted storm came around 4 pm, big gusts but all from the NW, so good for us. A little rain, maybe 5 mm. I feel much better. Maybe unreasonably so? Fire Info Officer said that we are entering into a warmer drier period. He emphasized that the fire season is not over. Half the Cariboo is still either on alert or on order.

I was very happy to see Pat and Talia when they arrived to spend the weekend with us; they offered to help fire-smart the place.

August 20.

> Calm and cool. Normally I wouldn't like to see the cool weather—fall coming on. But now I welcome it. The report on the Elephant Hill fire said there was activity on the northern and eastern flanks, but the fire didn't grow beyond its perimeter. There was some rain on the fire. Here we had about 1/2 inch. Feel much better than on Thursday when the smoke was heavy on the hills.

August 21.

> Eclipse. Sky got somewhat dimmer. Forgot to look at the sun. Felt a bit sad. Talia and Pat left. They were very helpful! We got the water tower painted with Lumber Guard and more wood split. But so much blowdown in June. Impenetrable tangles of downed trees—some quite massive. That's the real issue here. Climate change affects us in many ways—the pine beetle, blowdown and then hotter drier summers—a perfect storm. The worst fire season in the province's history.

August 22. The lake was quiet, unnervingly so. Hardly anyone was around. We went canoeing, saw maybe two cabins occupied to the east of us, but none to the west. I wrote:

> Gordon hosed down the place. How much good does that do? I still think we are too woodsy. How can we not be woodsy? I saw a toad yesterday. I liked seeing it. Fat with yellow markings. I often feel overwhelmed. And then I cook or burn brush in the fireplace. Feel a bit better.
>
> Gordon, true to form, is remarkably sanguine, cheerful. "You think," he says, "we're on the brink, on a precipice. But we're not. You think we're on the edge."
>
> "Yes," I say, "that's how I feel."
>
> Got smoky after lunch. Ash rained down on us. But we saw a few stars tonight. Maybe clearer. And I wished on one. "No fire." As if that does any good! The loons still call and we heard the sandhill cranes also.

After this, my diary goes silent. On August 23, our elderly friend Jane arrived for another visit, along with her younger sister, Nancy Platt. Jane stayed with us, and Nancy at her own place farther west along the lake. Jane was keen to come, but I wasn't sure it was a good idea. I always had in mind that we might have to flee at a moment's notice, and Jane wasn't exactly nimble. My son, Tom, and his wife, Leanna, also came for a short visit. While they were at the cabin, the smoke lifted and we had benign puffy white clouds instead. We swam and canoed, just like we used to. An update on August 25 showed that the fire had grown again—to 175,000 hectares.[76] The finger was still directed at us, but not significantly closer. Next day, the kids left and I wondered morosely whether we would ever be together at the cabin again.

The smoke was a putrid yellowish-grey when we woke on August 30. We couldn't see any distance down the lake, neither to the east nor the west. An acrid smell stung my nose. *This can't*

Chris Brown, owner of the Paradise Bay Resort at Sheridan Lake,
thought his place was doomed until the wind suddenly shifted.

be good, I thought to myself. I had a deep sense of foreboding. But I'd had that feeling before and nothing terrible had happened. It wasn't a reliable indicator.

Gordon and I had planned to drive to 100 Mile that morning to do some shopping. However, I worried about leaving Jane by herself in our boat-access-only cabin. While we had breakfast, I suggested, "Jane, how about we take you down to Nancy's while we're at the 100?" She flatly refused. I pointed out, "We'll have the speedboat. What will you do if an evacuation order comes?" She said, "I'll row out." The idea struck me as crazy. She was ninety-one and partially blind. "I'm not afraid." Jane said. "If I were afraid, I wouldn't be here." She was feisty, but that didn't mean she was *right*.

At that point, Gordon announced that he wanted to clear away some more brush and wet the place down before we left. My discussion with Jane was postponed. At 11:13 a.m., our friend Ann texted Gordon: "I think you should get ready because they are talking about evacuation south of the highway. I think it's happening sometime today." Ann's partner, Larry, works for Conair, a company that has a contract with the BC Wildfire Service to supply aerial firefighting. Her sources are good. Shortly after, Nancy phoned to say that she'd heard from Chris Brown at the Paradise Bay Resort at the western end of Sheridan Lake. He said the district was going to come to a decision very shortly. We rushed around putting away things that were strewn about the deck and dragging propane tanks down to the lakeshore.

"Should I strip my bed?" Jane asked. "Don't bother," I said. "If the place burns down, it won't matter. And if it doesn't, we'll be back." She laughed and went back to tidying up.

And then at 12:54 p.m., Gordon and I both got texts, phone messages and emails from the Cariboo Emergency Notification System, all saying the same thing: "Evacuation Order issued for an area south of Highway 24, including the areas of Watch Lake, Little Horse Lake, Little Green Lake and the western side of Sheridan Lake." We had registered on the system because without road access we assumed it was best not to count on the RCMP knocking on our door.

At least now, I thought, *I won't have to argue with Jane about her idea of rowing out.* We threw our clothes together. Gordon attached the fire hose to the pump and lashed it to our solar panels. Then he started the pump and directed the spray over the roof. The pump would run until the tank ran out of gas. It wasn't much protection, but it was some. I took a picture of Gordon on the roof setting up our "sprinkler system," sent it to our family, and labelled it "Gordon's Last Defence." Then we got in our boat and headed for the resort, where our car was parked. We met Nancy there; she was going to drive Jane home in her car.

Gordon and I had arranged to take our boat into 100 Mile for winter storage after Labour Day. We talked about whether we should haul it in a few days early or whether, to save time, we should just leave the boat tied up at the dock. "If we aren't running for our lives, we might as well take the boat to 100 Mile," I said. Gordon and I looked at each other. Were we running for our lives? We didn't know. We understood that if an evacuation order was issued, we were supposed to go—immediately. But how fast is that? Having crossed the lake, I felt a little safer. We decided we could chance it, and took the extra time needed to load the boat. I remember at one point peering past the resort dock into the sepia murk and seeing a water skier, of all things. Here we were wondering if we were running for our lives and someone else was water skiing! Talk about cognitive dissonance! But of course, only the south side of the lake was evacuated. The north side was still just on alert.

We had some trouble undoing a bolt on the trailer; I went to see if I could borrow a wrench from Chris. He found one in his shop. "Just leave it by the door when you go," he said.

"You're staying?" I asked.

"You betcha."

I wasn't surprised. I hugged him. "Be safe." I said.

CHRIS HAD SOLD THE RESORT in the spring but the new owners had been in no hurry to move in; they agreed to take possession in September and Chris stayed on during the summer. Brad Potter, who was the realtor on the deal, told me, "At the beginning of that fire, my buyer was phoning me and going, 'Brad, what's my situation if that fire gets as far as Sheridan Lake and burns down the resort? Do I still have to buy it?'"

Brad recalled saying, "Don't be stupid, that fire is so far away it will never *ever* get that far." Every week the buyer would phone and ask the same question, as the fire got closer and closer. And Brad kept saying the same thing. "No one ever imagined the fire would go over a hundred kilometres," he said. "One wind event took the fire from Watch Lake all the way up to Jack Frost Lake." As the crow flies, these two lakes are about six and a half kilometres apart. "The next day the winds came from the west and pushed the fire toward Sheridan Lake."

Chris was in a financial bind. If the property went up, the purchase agreement was toast too. Despite the evacuation order, he was determined to remain behind. He managed to persuade his wife to leave, but he was going to stay as long as he could, to protect the resort and his sale.

Chris had been hosing his place down for some time before the evacuation order was issued. When he first started, the pump he was using didn't have much pressure. The spray wouldn't go very far or very high. He figured he wasn't going to be able to fight the fire successfully with it. But as so often happened, there was a serendipitous intervention. Chris told me later, "The guy from the Honda shop pulls up in my yard with a brand new pump and suction hose. 'That's from Dave,' he says." Dave is another resident on the south shore of the lake, who also stayed during the fires. He has a big off-grid house with a large shop and numerous outbuildings. His nickname is Airplane Dave because he often comes and goes in his float plane.

I might have a chance now, Chris thought. He felt so encouraged that he bought a second new pump and set of hoses. He had both pumps running, hooking one up to his network of sprinklers and the other to the fire hose that he used to manually wet everything down. "I was soaking every tree as high as I could spray it. I'd work my way up and down, follow the branches, go to the next tree, the next tree, the next tree. I did that for a week on all these trees around the house as far as I could reach and as far back as I could go. I was even spraying down my wood piles, fire hose wide open on them. I thought, *If they catch on fire, I'll never get them out. Keep 'em soaked*. I bought sprinklers too. I had one in the backyard that sprayed all around and halfway up the trees. It would actually spray almost to the house. Another one I put on the trail here—it was spraying around in circles and on the house. One was going back and forth. The other was going 'round and 'round. Tens of thousands of gallons of water I sprayed around." Chris was using so much water that a low spot toward the back of his property turned into a small lake. But the inferno behind was snorting on regardless, engulfing the forest, spewing roiling ashy clouds high into the sky, and glowering with intensity.

Structural protection crews came on August 31. Chris had sprinklers on his house, but the crews gave him some more to put on the shop. The four redoubtable bombers from Quebec—241, 242, 243 and 246—arrived and Trevor Pugh, a local cottager, posted an impressive video on Facebook showing how they were hammering away at the fire: scoop, head for the target, drop and repeat, a five-minute cycle.[77] At this point, the fire was about two kilometres away from the resort (five kilometres from our place).[78] Even now as I revisit the video, watch the yellow planes, hear their engines drone and the loud swish as they swoop into the blue waters of Sheridan Lake for their load, I feel like punching the air with my fist and yelling *yes!* It's like looking at an old war movie: the fight is over, but you still want to root for your side.

The coordination of the attacking planes was lovely to behold, the responsibility of a BC Wildfire Service air attack officer who selected the targets and a bird dog pilot who together planned the run, led the bombers to meet the strike's objectives and made sure they were able to do so safely. Ryan Gahan is a pilot with Conair who spent thirty hours flying over the fires between Green and Sheridan Lakes in the summer of 2017. When I spoke to him in a coffee shop in February 2019, he said, "I'm the bird dog for the Conair Fire Boss group, but I work with everybody—Conair, Air Spray [from Alberta], the Quebec guys." He was an air traffic controller except that he didn't do the job sitting at a desk, but while flying himself. "The challenging part is having good situational awareness, knowing where all the aircraft are," he said. "Last summer I was on a fire at the Arrow Lakes just up from Castlegar. There's helicopters going up and down a ravine. I was working in the middle with skimmers and the retardant ships were coming overtop. We had ten aircraft on us. So it turned into this big, big dance."

To keep track of everybody, a bird dog plane has multiple radios; one is for communicating with the planes, another for talking to the ground crews, still another for connecting with the helicopters. When Ryan and an air attack officer work on a fire as big as the Elephant Hill fire, they may monitor two more radio frequencies, which provide a way of communicating across the whole incident. On top of all that, a traffic collision avoidance system displays radar on a screen. "It helps, but because there's a delay to it, it's not very accurate about where the planes actually are. But it gives you the number of aircraft that are out there." A big, big dance indeed.

ON SEPTEMBER 1, THE STRUCTURAL crews came again and set up sprinklers on Chris's cabins. "For a couple of days," Chris said, "we knew the fire was coming. No doubt about it. You could hear it roaring over there. I slept in the boat. The first night it was actually pretty quiet, except for the roar of the fire. The second night was terrible. By then they had pumps all over the lake. They are noisy two-strokers and they run wide-open, very little mufflers on them. Then I heard someone off-load a Cat. And they're running around with that. Squeak, squeak, squeak. Clank, clank, clank. Pumps were roaring. It was cold. The second night I didn't get much sleep."

Despite the efforts of the hard-working pilots, the fire came through about 4:00 in the afternoon on September 2. "I just watched trees burst into flames. There was absolutely no doubt it was coming," Chris said. "You could see the flames—hundred-foot flames coming off the trees. By the time it jumped the little channel, the little inlet or whatever, all these guys said, 'We're bucking out of here. We're gone.' Animals couldn't outrun it, it was coming through that bush so fast. Maybe running straight down a road they could, but jumping over logs, they couldn't. The wind was doing 40 K. The fire was leaping, blowing. It come in here with a vengeance."

Carson Dorward was in charge of one of the structural protection crews working at Sheridan. He used to work full-time for the BC Wildfire Service and at one time owned a ranch at the southeast corner of Sheridan Lake, so he was familiar with the area. When I talked to him on the phone, he said, "If a fire is really rolling, standing in front of it is not going to do any good. When you've got a fire going through the treetops, you can put a bunch of water down in front of it. It may skip a little bit but the heat that it's putting out will evaporate the water. The sparks will jump over what you've done. You can go along the sides of it and prevent it from widening out. As far as stopping the head of a fire that's actually rolling, Mother Nature's the only one that's got the secret to doing that."

One small dirt road led out of the resort to Magnussen Road and then to Highway 24. If fire engulfed that narrow track, anyone on the resort property would be trapped. Carson didn't want to take that chance: "Leaving people in a situation where they are going to be surrounded by fire is not something we can justify doing. We made the decision to make a tactical withdrawal and to continue working in the rest of the subdivision until we could be sure it was safe. You can never eliminate the risk. But you can mitigate it."

All the ground crews were pulled off the fire. The air crews left as well because it had become too smoky for them to operate. However, Carson made Chris a handsome offer. He was willing to stay behind: "If you want, I'll come out in the boat with you. When this thing blows over, we'll come back and put out spot fires." They jumped into Chris's boat moored at the dock, went out into the lake and anchored offshore where they could watch the unholy spectacle unfolding on land. Chris told me, "It came behind the resort here. The speed that fire was coming through, I didn't think we'd be there long. We sat out there, and the lake was really rough—noisy and rough. It was crazy. Smoke just rolled through. This whole place disappeared."

Ryan Gahan, a bird dog pilot with Conair, spent thirty hours flying between Green and Sheridan Lake in the summer of 2017.

Meanwhile, Gordon and I were at home in North Vancouver, wondering what was happening at the lake. We consulted the official sources: the Emergency Operations Centres and the Fire Centres. But sometimes we wanted more information than they provided. My daughter put me onto a YouTube channel called *brents desk*.[79] Brent posted regular updates three or four times a day, sometimes more often than that. On his channel, he had links to BC, Canadian and US sources—over twenty sites that provided fire map data, wind and weather data, webcam images and, most engrossing of all, hot spot data. Satellite sensors seven hundred kilometres up in space can detect a backyard fire that is only 3 × 3 metres.[80] They are sensitive to objects hotter than about 200°C and since wildfires may reach temperatures of between 800°C and 1,200°C,[81] they are eminently detectable. If smoke or clouds obscure the view, however, maps of hot spots might under-report the real activity.

Brent walked viewers through the information, explained what was of interest, and cautioned about inaccuracies. He reminded his audience that hot spot data could be off by half a kilometre or more. He advised those needing to make a decision to use more than one source and to always pay attention to boots-on-the-ground reports, if available. I liked his careful approach and the way he provided both data and education so members of the public could better understand things for themselves.

Though Brent used a YouTube channel, we only heard him and never saw his face, only the maps and images he was talking about. He never mentioned his last name. He was like a mysterious war correspondent. I grew to depend on him and I would instantly recognize his opening "Greetings" anywhere. Brent became our lifeline. I checked his channel first thing in the morning and at night before I went to bed.

Other people felt the same way about *brents desk*. Cathy Smith, whose cabin at Pressy Lake luckily did *not* burn, told me: "Everybody was glued to *brents desk*. Anytime I got a new post from him, I shared it on our Pressy Lake page. A little while ago they were asking people to put people's names forward for unsung heroes. I tried to put him forward for recognition, but he's like that Carlton the doorman [heard via the intercom but almost never seen in the seventies sitcom *Rhoda*], a mystery guy, and I didn't have his email, his address or his name even." Alicia Polanski, whose story I told in chapter 20, was also a cottager at Pressy Lake. "I love Brent," she said. "I don't know who in hell he is. I think his name might be Kyle. In all honesty, he's the person who kept a lot of us sane. Everything was facts. You could not get facts from anybody. Everybody gave you the runaround."

Later, I confirmed that Brent was not Kyle, but really Brent Lewis. He had a farm in Port Coquitlam and a place in Chasm that he described in an email as "one of our most special places to be."

The information available from official sources was not detailed enough for him, he said. "I was beginning to understand the fire was dynamic and changed positions without notice. I needed to see, hourly or better, real-time information on where the fire was." A search led him to the National Research Council Interactive Infrared Map, which told him where the new fires were breaking out. He did a lot of reading online to understand what he was looking at; he began to grasp the importance of wind direction and discovered a website called Windy.com. "When I found the information on wildfire infrared and wind helpful, it just felt like the neighbourly thing to do was share it." From July 20 to September 28, he posted 150 videos.

In my living room in North Vancouver, looking for news on the fire, I turned to Brent's most recent post. "Greetings, it is September 2 at 5:00 p.m. Let's go to Windy right now and see the pattern that's occurring." He directed the viewers' attention to a spot just above the fire line where winds from the north were meeting winds from the south. "Essentially," he said, "it's the wind rotating around and it may actually blow the fire back in on itself and that would be ideal. Let's take a look at the infrared now. This is the VIIRS system and we're showing data from 4:00 p.m. Pacific Standard Time and as you can see a few concentrations of new infrared on the fringe area ... "

Two hours later: "Greetings, this is September 2 at 7:00 p.m. and this is the issue of the day: that smoke is obscuring much of the visibility and it could be obscuring infrared data. As you see on this screen, we see a few random dots by the Modis system but if we look at the photographic material provided by the Beddow Tree webcam looking west at Sheridan, we see smoke in the distance. So we're left with two big questions: how recent is the data and how accurate is the data?"

And then four hours later: "Greetings, this is September 2 at 11:00 p.m. If you look to right of centre of your screen you'll see Sheridan Lake, and southwest of that is Number Two Lake, and we can see infrared being displayed on the southern shore of that lake. I've heard unverified reports that activity was occurring today within the last twelve hours near to Paradise Bay at the southwestern end of Sheridan Lake ... "

Those reports may have been "unverified," but they were true nonetheless. "The smoke was coming closer and closer to us," Chris told me, as he continued to describe what he and Carson were seeing as they were moored offshore. "I didn't know how hot it was, but it was solid black. It wouldn't have been good to be in that. I was just thinking maybe I'd better move back. I was about to start up the boat and turn around when all of a sudden you see that cloud of smoke clear away. The fire quit moving to the east. It just petered out and started going to the south."

Carson said, "I remember looking at it first and going, 'The wind's shifted, but it will come right back again.' Then I was like, 'The wind's shifting. It's shifting!' A set of goosebumps comes. 'It's actually pushing it away. The smoke is starting to clear.' Now you know it's not just a gust in the other direction, but a shift. It was pretty cool."

Don Schwartz, another Sheridan Lake cottager, had captured some of these moments in a short video he posted on Facebook. You can't see the resort—it is enveloped in a thick blanket of smoke that reaches to the water's edge—but the fire is clearly moving at a good clip.[82]

When the smoke lifted, Chris and Carson came back to land and started looking over the property. "I figured there'd be nothing left," Chris said. "I still thought probably everything was on fire. I wasn't relieved yet. I was still pretty shook up. I landed on the little dock straight out front. I come 'round, I could see the house was standing. I was still in disbelief that nothing was on fire. After we walked through the backyard, I went, 'I just can't believe it; not even that little shack was burnt.'" A couple of old rotten stumps were smouldering, but Chris's house, the shop, the six cabins and the boathouse had all survived. The fire had come right to the edge of his property line but had not crossed it.

As soon as Mother Nature had done her bit, the firefighters mobilized to take advantage. The crews took a boat to our friend Ann's cabin. Her cottage was the closest to the resort at the southwest corner of the lake and the fire came to within fifty metres of it. The structural protection crews put sprinklers on her roof and on Nancy's, which was next door.

"Even after it went south," Chris said, "it was still behind all the houses. I spent four days with the firefighters, ferrying them, sprinklers, hoses and pumps to all the cabins." All along the south shore, every building and shop was outfitted with sprinklers. Chris said, "They did not miss much. Guys would work until dark. I'd go pick 'em up. They'd be over there flashing their lights so I could find them. We ended up changing out a pump at 10:00 at night. I went to pick those guys up and they couldn't get the pump started. It was the closest place to the fire line. They tried and tried. I said, 'Okay, let's go and get another pump.' We set up a new pump, started her up." The pumps were hooked up to double tanks so they could burn about thirty-eight litres of gas and run for nearly twenty-four hours without being refilled. They were linked to hose networks, so each pump generally served three properties. Chris said the crews were bringing in nearly 570 litres of gas a day, enough to provide structural protection for around forty-five places.

I had the impression that the BC Wildfire Service was determined to slay the dragon once and for all and pulling out all the stops. The air support continued, now from BC's Conair fleet and helicopters—Bell 214s, Bell 205s, A-Stars—bucketing up to thirty-eight hundred litres of water. Chris thought probably a half dozen to a dozen heavy equipment operators were working on a fireguard in the back. They wanted to give the cabins on the south side some protection in case the wind shifted again and blew from the south.

The first indication I had about the probable fate of our cabin was a post on the Sheridan Lake Facebook site. On September 3 at 10:10 a.m., Ann's sister Martha wrote:

> I spoke with Cindy this morning. They went out in the boat and checked all our cabins and everything is still okay. She will give a full update when she gets home tonight.

The Elephant Hill fire had grown to 192,000 hectares.[83] The bloated monster died just one kilometre away from our beloved place. Like Chris, having expected disaster for so long, I felt strange when the danger passed.

CHAPTER 24

YOU GET USED TO RIDING WITH FLAMES

ON SEPTEMBER 12, GUS HORN, a rancher who lives close to 100 Mile House, posted a video of what he saw when he flew over a number of nearby lakes with a couple of friends. Sheridan Lake, Watch Lake, Green Lake, Thomas Lake, Jack Frost Lake and Pressy Lake all unfolded below him. There was no voice-over, just the steady thrum of the engine. I found it heartbreaking to see the scorched land, the black lifeless stumps. Great swaths where there wasn't even a blade of grass, just the naked scarred earth burned to a light grey colour. Without its normal green cloak, you could see every crevice and hummock. The land's contours were all exposed. I felt its vulnerability.

I met Gus on his own place, which was untouched by fire. A busy man of broad interests, he was also the proprietor of Critical Mass Pop-Up Gallery, a venue in 100 Mile House for shows of photography, paintings, musical performances and films. We had a hard time arranging when to meet; Gus didn't want to take time away from work to talk to me and suggested that I interview him while he was doing something else. His first idea was that I should accompany him on a trip to take some of his cattle to an abattoir. The timing didn't work out, so in the end our discussion took place during what was for me at least a more conventional errand, while he drove into town to gas up and do some other errands. We rattled along in his truck with his faithful dog, Torch, sitting quietly

Gus Horn, seen with Torch on the family ranch near 100 Mile House, got used to riding through flames.

behind us. Only after his chores were done did Gus allow himself to sit down in his living room. I also met Gus's mom, Helen, a spry ninety-three-old. (Actually, "spry" hardly does her justice. For her ninety-third birthday, she went whitewater rafting on the Fraser River for nine days, from Soda Creek to Yale, and pronounced it "the trip of a lifetime.") While we sipped the homemade smoothies Helen served us, I noticed that multi-tasking seemed ingrained in the Horn way of life. Their furniture was like that too: a dog cage doubled as a coffee table.

Gus's grandparents homesteaded near Roe Lake around 1912. In 1947, his father bought the ranch just east of 100 Mile that Gus now runs. It turned out that we shared a connection. Gus's great uncle was Carl Nath, who settled on the north side of Sheridan Lake and was helpful to Gordon's mom when she and our friend Jane bought property on the south side. Carl was the pioneer who walked from Ashcroft to Sheridan, using the same route that the Elephant Hill fire later followed. He and another early settler, whom I know as Old Man McNulty, built our boathouse. In our cabin, we have a photograph of Carl and his wife, Sarah—the black-and-white picture hand-tinted in soft pastels.

In September, Gus went riding with a friend, Ron Eden, to help drive the cattle Ron had in the fire zone south of Sheridan Lake. (Ron and Gus go back a long way: Ron's father was the best man when Gus's father got married.) About a dozen of Ron's cattle had burned to death, but as he and Gus rode through the countryside, still pocked by fire and smoke, they discovered that most of Ron's animals had managed to save themselves. "Cattle at that time of year, especially yearlings, are in little batches," Gus said. "You've got three head or five head or six head together. Occasionally there's one by itself. You know where the water holes are and they find water regularly. They move around all the time, but it's not that they're trying to escape from you." Ron and Gus set up temporary corrals behind the lines, brought the cattle they found to the nearest set, and then transported them out with a truck. Gus also took videos and pictures while driving cattle through the smouldering rangeland.

"The fire was moving slow enough while I experienced it. If it was burning hot in the bush, if it was candling, usually that was in the evening, when things warmed up during the day. You might be riding or driving on a road, you're a hundred yards away or maybe even closer than that. There were lots of clear-cuts, lots of open areas. It will burn across those areas. You get used to riding with flames." Gus and Ron kept an eye out for trees at risk. "Certainly we took it upon ourselves to cut lots of trees off the road. But if there is a tree fallen on the road and you're cutting one off, there's a chance that another one might come down. You don't want to be there when it's windy, no doubt about that." And they were always conscious that the full-throated blaze was not far away. Gus recounted, "At one point, a conservation officer said, 'Turn off your vehicle.' So I did. He said, 'Hear that?' We were close enough, you could hear the fire. It was a roar. It was really heavy. You're aware and you're not going to hang around there."

Gus spent two weeks in The Zone. "It was horrible. You have to be careful, it's dangerous," he said, "but after coming out, everything seems so banal. After such a big crisis, it's hard to get used to the day-to-day. I had a buddy who was in Vietnam, who went back for a second tour. I couldn't see why, but now I have a glimmer of understanding." Gus had hit upon something I had not thought about much—that sense of excitement you got when being in The Zone, just being away from the "day-to-day." Jeremy Vogt, the pastor of the Cariboo Bethel Church in Williams Lake, had mentioned something similar. "You know, Claudia, some of it was just fun. I mean the empty city. The mill was shut down. I was the only one in my house. I was the only one on my block and it was just dead quiet. One of my friends had parked their big motorcycle in my garage so I took it out for a drive. Big empty streets, you know? All the RCMP were out of town." He chuckled at the memory and I smiled too. I had this image of a pastor on a Harley gleefully roaring through Williams Lake, robes flying, and no one around to check him.

ATTITUDE MADE SUCH A DIFFERENCE. As the poet Charles Bukowski said, "What matters most is how well you walk through the fire." I thought back to the people I'd met and how brave some of them had been during the crisis. Cory Dyck calmly looked at the conflagration heading for the Becher Ranch and said, "We're ready for that." This was both admirable and necessary. As Blake Chipman remarked, "In firefighting, if you panic, you die, or you cause somebody else to die." Heather Gorrell drove through flames to help a woman rescue her horses at a place north of Williams Lake. It was so dangerous the RCMP who had been manning a checkpoint on the road to that property had abandoned it. But Heather suppressed her fears and helped the people around her.

The individuals I spoke to were resilient, resourceful and hard-working. If they had not been, the outcomes might very well have been worse. The community of 16 Mile House is too small to fund an official fire hall. Still, the residents took it upon themselves to prepare for the possibility of fire. They built trailers capable of carrying from one thousand to six thousand litres of water, an innovation that helped them save the Woodburns' ranch. Dean Miller at the Chilco Ranch had his own excavators, cats, low-beds, water tender, fire pumps and thousands of feet of hose. When the fire came, he and his grandsons were ready.

Many of BC's firefighters went all out in their battles during the summer of 2017. Glen Burgess was deployed as an incident commander for eighty days during the summer of 2017. "A lot of those nights I'm getting three hours sleep, maybe four." He worked fourteen-day tours and although he was supposed to get three or four days off between deployments, often that didn't happen. Some firefighters went for long stretches without any sleep whatsoever. The smokejumpers on the West Fraser Road fire stayed up for thirty-six hours straight to make sure the blaze they were attacking was properly contained. Members of the Blackwater Unit from Quesnel were up all night while the fires threatened Williams Lake Airport.

Not only were the days long, the fires were long as well, which affected the self-esteem of the crews. But they stayed on the job nonetheless. "It's hard when we're not succeeding because we are all people used to succeeding," Glen Burgess said. "A fire season like this where we weren't successful every single time was very demoralizing. One of our crew members was on the Elephant Hill fire for six or seven rotations. That's unheard of for these guys. They're used to going in, kicking some butt, getting done and moving on.

"Emotionally there's a big toll on us," Glen continued. "I still think about [the] Rock Creek fire, for example, a couple of years ago. It was my team that was there, you know. That was something. To get to where we were set up, we had to drive through the fire. I think I'm pretty good at suppressing emotions. But the next spring, I drove through there to visit a friend of mine. I had to stop, it hit me so hard. So you know there's that lasting impact." Glen reminded me that some of the worst scars the fires left were the invisible ones.

DURING THE SUMMER OF 2017, the province relied on a mix of paid workers and volunteers to fight the fires and assist those affected by them. When BC declares a state of emergency, the Office of the Fire Commissioner gains the ability to send resources—personnel and equipment—anywhere they are needed. In 2017, firefighters who were sent out of their own communities were paid either what they would earn at home or $41 an hour, whichever was greater.[84] Many communities pay firefighters on an "on call" basis—that is, only if they are actually working on a fire. If they are paid by their local municipality or district, they typically earn considerably less than $41 an hour. Golden fire chief Dave Balding, who writes about firefighting issues, said in a phone interview that in his town, "firefighters get between $13 and $18 an hour, depending on their qualifications. And some communities still have true volunteers that get no money whatsoever." In Ashcroft, when volunteer regular firefighters put in long hours during the summer of 2017, they earned less than the wage of $7 an hour for hot, dangerous work. They went way beyond the call of duty.

I talked to people who were deeply committed to supporting their neighbours. Lyn Arikado codirected emergency support services in Kamloops that ran the evacuation centre there. Herself a volunteer, she started with a core group of thirty-six volunteers when the fires broke out on July 7. By the end of the summer, 1,540 more people had signed up to assist. The Kamloops Reception Centre provided a total of thirty-eight thousand hours of volunteer services to over eleven thousand evacuees. Also in Kamloops, volunteers spent nearly thirty thousand hours caring for animals that had to be relocated due to the wildfires. Many organizations, agencies and individuals pitched in.[85] The Thompson-Nicola Regional District estimates that over the whole region, volunteers working on their own or with organizations spent over 150,000 hours to help people and animals affected by the fires.[86]

Val Severin, a manager of the South Cariboo Search and Rescue, is not paid at all for her work. Assisting with evacuations took up so much time she had to take leave from her regular job. When I asked her whether she got anything from her employer during the summer, she said, "A bit. I did receive some wages. My employer was awesome. Many businesses did get some insurance for loss of business, so they were able to share that with their employees." But as much as she appreciated what her employer did for her, she is still out of pocket due to lost wages.

I was pleased to see that in their 2018 report, George Abbott and Maureen Chapman suggested compensating people who house evacuees for any expenses and paying search and rescue "volunteers."[87] But I think we need to revisit our reliance on volunteers in general—especially if mega-fires are becoming more common. Volunteers may be able to help out for one season or two. But year after year? It hardly seems fair or sustainable. Roger Hollander, the fire chief at 100 Mile House, also has concerns about depending on volunteer firefighters. "Gone are the days of the old bucket brigade. We're no longer passing buckets to each other to throw on the neighbour's house. We're at the level of a professional. Would you ask your doctor or your surgeon to volunteer? Do you think the pilot who is flying you to your next vacation is a volunteer? What I'm seeing is what the fire services North America-wide are seeing: there is a drop in volunteering." This is due in part to the fact that the fire service is demanding more of its volunteers, both in terms of the level of training required and the number of calls to which they are asked to respond. Roger told me that the 100 Mile House department used to get "twenty-five, fifty, maybe a hundred calls a year." In 2017, it responded to 450. "Obviously there's only so many tax dollars to go around but, boy, something has to change if you're going to sustain a professional service, because we don't ask that from any other group," Roger said.

While the volunteer firefighters that Roger mentioned were part of an organization, people often saw a need and took it upon themselves to fill it quite spontaneously. In Ashcroft, for about a week Heather Aie and other volunteers organized meals for the firefighters and members of the community who weren't able to cook when power went out in the village. They served up hamburgers, spaghetti dinners, pizzas, breakfast sandwiches, pancakes and sandwiches. They got donations, which they distributed: water, baked goods and fruit from a local roadside stand. "More and more people came trickling in to help. They wanted a purpose and something to do," said Heather. "It just snowballed." When an evacuation order was issued for Clinton, Jin Kim, the proprietor of Clinton Shell Gas and Budget Foods, dropped his gas prices to help customers who were leaving. Pam Jim, who owns Jim's Food

Catherine Clinckemaillie, of Skookumhorse Ranch near Clinton, had to get horses, yaks, goats, sheep, chickens, dogs and cats to safety.

Market in Little Fort, gave away sandwiches to hungry evacuees. Many other businesses, large and small, also donated food, services and supplies.

And as we have seen, volunteers helped animals as well as people. Lana Shields coordinated the rescue of over three hundred horses from Williams Lake and surrounding areas. Most were transported to Prince George over a period of five days, an effort undertaken completely by volunteers. On July 7, Dawn Bigg, who lives in Horsefly, took in a family of four evacuees. She accommodated not only a mom, dad and two kids, but also their three horses, one cow, three or four goats, one sheep, two cats and two dogs, as well as two pet rats. "I didn't find out about the rats for three days," she said with a laugh. This meant, of course, getting them to fit in with what she termed "our own crew"—fourteen pigs, twenty-four chickens and three dogs. A week later Dawn welcomed another family, who arrived with three ducks, nine chickens, two dogs and two cats. The family also had six cows for which Dawn found another shelter. Catherine Clinckemaillie, who owns

Malcolm James, a rancher in Canim Lake, found asylum for
his horses, cows and llamas in the Columbia Valley.

the Skookumhorse Ranch north of Clinton, had twenty horses (some were hers and some were boarded with her), fifteen yaks, twenty Markhor goats, sheep, chickens, dogs and cats. When she was evacuated, the Cherry Creek Ranch near Kamloops gave her menagerie a refuge. Malcolm James, who lives near Canim Lake, had four horses and twelve cows, as well as dogs and llamas. He found a safe place for all his animals with friends in the Columbia Valley, near Cultus Lake.

No humans died as a result of the fires. One way or another, officially or unofficially, people were notified about evacuation orders. Access to our property at Sheridan Lake was by boat only and we figured no one was going to knock on our door, so we registered with the Cariboo Emergency Notification System. As we had surmised, when the evacuation order was issued, no one came to our place. But Gordon and I both got texts, phone messages *and* emails. Evacuating people in the Interior of BC can be complicated because some of the homes are remote, on back roads and out of cell phone range. Two hundred and fifty people were caught in Bowron Lake

Provincial Park when that aggressive wildfire started. Park rangers and operators stayed up until midnight to make sure everyone was informed and safe. Despite the smoke and the wind, no boaters capsized. Raylene Poffenroth in Riske Creek recalled that her husband, Bryan, was building a fireguard when he told her that he was concerned about a couple of neighbours—Mike and Connie Jasper. "I go to their house and there's nobody home," Raylene said. "So I just go in the house and call Connie at work and say, 'The fire is coming. What do you need? You've got to get out.' I started evacuating some of her stuff and bringing it down here." Cultural factors added complexity. When I talked to Francis Laceese, Chief of the Toosey Nation in Riske Creek, about the evacuation of his people, he said, "Some of the elders don't understand English. Tsilhqot'in is their first [only] language." So when the Toosey Nation signed an evacuation order, the police went door to door accompanied by a council member who spoke Tsilhqot'in, to ensure that everyone understood what was happening.

From newscasts, articles, Facebook posts and conversations, I heard and read again and again that the fires of 2017, though unprecedented in ferocity, were not an anomaly. A report by Canadian federal scientists, published in April 2018, warned that by the end of the century large parts of Canada's boreal forests could die due to disturbances brought about by climate change—fire, mountain pine beetle, spruce budworm and drought.[88] Werner Kurz, a senior scientist with the Pacific Fire Centre at the Natural Resources Canada office in Victoria, put the point to me this way: "We are on a trajectory of increasing climate warming, greater drought and higher temperatures. What we saw in 2017 and 2018 were stepping stones, or steps on a trajectory, for even greater fire risks in the future."

The folks in the Cariboo will step up to the plate to cope with changing conditions, as they have always done. They will fire-smart their places. They will buy pumps, hoses and sprinklers. In so doing, they will defend not only themselves but their neighbours. But I am afraid that unless we *all* do our part, these efforts will be for naught. We need to grapple with the fundamentals, re-examine what stewardship of the land really entails and revise our forest management practices.

We also need to address the climate crisis. This is often framed as an exercise of denial: we are told we must give up flying, using plastic, eating meat and dairy, and so on. Such a future can seem small, cramped—not very appealing. But there are rewards in rising to an existential challenge: being swept up in a cause greater than yourself can be deeply satisfying and enriching. As pastor Jeremy Vogt said to me, "It was an amazing time. I just reflect on it in many ways as a positive event for our community because it brought us together." The RCMP's Svend Nielsen said something similar: "I don't want it to happen anywhere else, but just for that feeling, part of me wants to have it again. It's that camaraderie."

As British Columbians demonstrated in the summer of 2017, remarkable things can happen if we take responsibility, come together and act.

CHAPTER 25

LET US CONSPIRE WITH THE FORESTS

AFTER THE FIRE DRAGONS RAMPAGED across the province in 2017, they returned the following summer. Even more land was burned—1.35 million hectares—but 2017 remains the second worst year for fires and still holds the record for the largest number of people driven from their homes: sixty-two thousand.

The costs of suppression alone totalled $568 million, making 2017 the most expensive fire year in BC's history.[89] And as Robert Gray, a fire ecologist based in Chilliwack, points out, the indirect and additional costs of fires of this magnitude can run up to thirty times the direct costs of fire suppression.[90]

BC lost 53 million cubic metres of saleable timber,[91] equivalent to one year's allowable cut, worth somewhere between $6 and $7 billion.[92] Eleven wood-processing mills in the Cariboo-Chilcotin closed during the fires at a cost of $2 million per facility per day of shutdown.[93] Three hundred buildings burned. The provincial cattle industry lost over four hundred thousand hectares of rangeland (approximately a million acres). Thousands of kilometres of fences were incinerated. Many of the thirty-five thousand cattle in the fire-affected regions went missing.[94] The province spent $6.9 million a day to cover travel, shelter and food expenses for the evacuees.[95] Smoke from the Interior drifted down to the Lower Mainland and caused emergency room visits to spike by more than 20 percent as people sought help for respiratory issues.⁹⁶ In the spring of 2018, floods developed because fire-ravaged lands were not able to contain rain and melting snow. The floods forced five thousand people to leave their homes again.⁹⁷

A report by Canadian scientists at Environment and Climate Change Canada and the University of Victoria, published in December 2018, concludes that human-caused climate change has increased the risk of catastrophic wildfires in British Columbia.⁹⁸ This is not something we can easily fix. Nevertheless, according to Ray Travers, a forestry consultant who lives in Victoria, we *can* improve the resilience of our forests. In a long phone conversation, he suggested I look at Sweden for inspiration.

Sweden has about 40 million hectares of forest, British Columbia 60 million. Ray told me that they have the same amount of commercially viable forest land—22 million hectares—sometimes called "managed forest."⁹⁹ In both places, people have deep feelings about trees. "The Swedish constitution grants Swedes *allemansrätt,* 'the freedom to roam.' It gives all members of the public free access to nature and wilderness, including forests and water, even though most Swedish forests are owned either privately or by companies."¹⁰⁰ In British Columbia, in 1993, twelve thousand people protested logging in Clayoquot Sound. A thousand people were arrested but ultimately the protesters were successful in protecting the old-growth forests.¹⁰¹ I myself treasure memories of visiting Meares Island, a Clayoquot Sound treasure, home to some of British Columbia's most ancient trees. Over a thousand years old, they are impossibly gnarled and twisted, the very epitome of "old," and yet they sprout fresh needles every spring and flaunt an array of bright green ferns in their crevices.

Since the turn of the last century, the Swedes have been thinking about how to preserve their forest resources. In 1903, Sweden passed its first modern forestry act to secure the supply of wood. Between 1950 and today, the standing volume of Sweden's trees has tripled from 1 billion to 3 billion cubic metres. Sweden achieved this despite the fact that it cuts more timber than we do—85 million cubic metres annually.¹⁰² The thriving industry employs a hundred thousand people.¹⁰³

When logging began in BC in the late nineteenth century, environmental concerns, species protection and preserving something for our children and grandchildren were not top of mind. The annual volume of timber harvested in the province increased from 1900 on, peaking at almost 90 million cubic metres in 1987. In 2010, it dropped to 50 million cubic metres, due to the pine beetle epidemic. By 2015, it had recovered some, but the harvest is not expected to return to the heady days of 1987 ever again.[104]

Sawmills were shuttered due to a lack of timber supply even before the fires of 2017 and 2018. "Between 1990 and 2015, the number of large- and medium-size sawmills in BC declined from 131 to 70,"[105] Bob Williams, a minister in both the Barrett and Harcourt governments, wrote in 2018. When I spoke with Ray Travers, he said that he expects more mills and companies to go. West Fraser, which has twenty-four mills in BC, also has thirty-eight in the US and recently bought a dozen more there. "They're leaving," Ray said. "It's not surprising at all when all you've got left is wood that's financially not viable—on steep ground or remote." No wonder the number of people directly employed in our forests has dropped from eighty-five thousand in 1997 to sixty thousand in 2016.[106] And as Ray predicted, in 2019 more mills announced closures and indefinite curtailments. Vavenby, Quesnel, Kelowna, Chasm, Prince George, Mackenzie, Maple Ridge, 70 Mile House and Fort St. James were just some of the places affected.

We can learn much from Sweden's careful stewardship—the way it extracts value at every stage of the lifespan of a forest. There, stands are periodically thinned to increase the yield and create a higher-value product. The "thinnings" are not discarded but processed in various ways. Ray says they are the source of 30 percent of Sweden's wood. When mature trees are finally harvested, the branches and tops are used by the biomass industry, which turns them into chips and pellets, mostly used to generate heat. Nothing is wasted. In BC, we don't thin our trees, nor do we make much of an attempt to use the branches left on the ground after a tree is felled. Sweden enjoys a revenue stream we don't have.

Swedish forests have been resilient in the face of fires. Like BC, Sweden had a very hot and dry summer in 2018. People were evacuated and cows died because of the drought. But here's the interesting part: only twenty-five thousand hectares burned. "This is far more than we are used to," writes Torbjörn Johnsen.[107] Yet the amount is dwarfed by the comparison to what went up in smoke in BC—over a million hectares.

In Sweden, 50 percent of the forest land used for logging is owned by individuals and 25 percent by private-sector companies. (Churches, non-profits, the state, state-owned firms, and local and county councils own the rest.) The individual holdings are often small, around fifty hectares, and frequently part of multi-generational farms.[108] In BC, the Crown owns 95 percent of our province. "That's a big difference between Canada and Sweden. When you're operating over there, you're operating on the land your grandparents operated on," Ray Travers said. It may be that in Sweden private forests are husbanded more carefully than our publicly owned Crown land. It may also be that the small woodlots attached to family farms with pastures have a beneficial patchwork quality.

But I kept thinking about Gary Filmon's report and the recommendation to remove brush from the forest floor. In twelve years, the ministry dealt only with eighty thousand hectares, 8 percent of the problem. However, in Sweden, a quarter of Sweden's energy needs are met with biomass,[109] largely derived from logging residue.[110] The waste has value, so the Swedes have no problem collecting it, thereby reducing a fire hazard.

BC already uses some waste wood by turning it into pellets that can be burned; 90 to 95 percent comes from the scraps and sawdust that accumulate at mills.[111] Very little comes from the thinning of trees or from the brush and limbs left over from logging. Gordon Murray, the executive director of the Wood Pellet Association of Canada, would like to change that. "We rely on trying to buy it [wood waste] from the logging companies after they finish harvesting. They just burn it. We're out there just begging the government to give us access to it, to private companies to give us access to it. And, you know, all this burning that the government is now saying that they're going to be doing. I mean—this is a crime, an absolute crime. They're burning up material that our industry could use."

In the US, the Little Hoover Commission, a bipartisan oversight group that advises the California legislature, recommended turning to the biomass industry to help prevent fires in California.[112] I asked Jens Wieting, a senior forest and climate campaigner for Sierra Club BC, whether we should do this too. "It all depends," he said, "on how much is being removed and what it is. A natural forest always has some dead timber and woody debris and that's very important for insects, for birds. It should not be removed. But if there's logging and some wood waste, it makes sense to use part of it for local energy use. Some removal of biomass from the forest is appropriate in moderation and if it is carefully planned. Some areas should be kept cleaner to reduce the risk of fire. But it must be done carefully and not across a vast area."

Rethinking our approach to biomass is one way of making our forests less flammable. Robert Gray, the fire ecologist I mentioned earlier, is trying another approach: studying a 1.3-million-hectare tract of forest near Quesnel to help us learn from the past. "There's actually a group of us in the scientific community who are looking at restoring landscape resilience," he explained in a phone conversation. "What we find is that historically on those landscapes, especially out in the Cariboo or Chilcotin, there were a lot of disturbances but they were small. Because of that, we didn't have these big blowouts because the landscape was much less vulnerable, not only to wildfires, but also to insects.

"We had a lot of recovering forests that didn't have a lot of fuels, a lot of deciduous hardwoods, a lot of grasslands, and a lot of areas that had burned multiple times and that just didn't have any fuel on them. A good chunk of that landscape was really in a kind of a basically non-flammable state. And we exclude fire for a century and now we have a landscape that's primed to carry fire for distances. We have to adopt those historic patterns but also keep an eye to climate change and restore a lot of those patches—those basically non-burnable or low-flame patches. Some of that involves prescribed burning. Some of it is conversion to hardwoods." (Pines are eight times more flammable than deciduous trees.[113] Aspens, on the other hand, make such good firebreaks they are sometimes referred to as "asbestos forests.")

According to Robert, "The research tells us that upwards of 40 percent of that landscape needs to be in these really low-flammable states." Once the restoration is achieved, he believes we will see the size and severity of fires decrease. "That's what we're after: more resilient landscapes."

FORESTS ARE THE LIFEBLOOD OF many BC communities. They are also at the front line of our fight against global warming: "I often say we cannot save the climate without saving the forest and we cannot save the forests without saving the climate," Jens Wieting said. A mature healthy forest can inhale tonnes and tonnes of CO_2 and sequester it safely for centuries. Not too long ago, BC forests were a major carbon sink, but they have now become a source of CO_2—a disturbing development.

Jens showed me a provincial government report where I could see this story unfold.[114] In 1990, BC's forests inhaled 101 million tonnes of CO_2 more than they released because of decaying leaves and rotting wood. Wildfires, slash burning and decomposing forest products (like paper, pallets) added 41 million tonnes of CO_2 to the atmosphere. The net effect was 60 million fewer tonnes of CO_2 in the atmosphere. This fully compensated for the 56 million tonnes BC residents emitted through burning fossil fuels, industrial processes and agriculture.[115] Not so long ago, thanks to our metabolically active healthy forests, BC was carbon neutral—actually carbon negative.

From 1990 to 2000, our forests faithfully absorbed over 100 million tonnes of CO_2 every year. But in 2001, as the effects of the pine beetle epidemic began to be felt, those great lungs started to falter. In 2003, BC forests sucked in 81 million tonnes of CO_2. But goosed by the wildfires, forest emissions added 5 million tonnes of CO_2 to the atmosphere. Our woods shifted from a carbon sink to a carbon source.

And then there was 2017, when the wildfires pumped out a record-breaking 177 million tonnes of CO_2. If we take into account our other forest-related absorptions and emissions for 2017, we see that the *total* emissions coming from this sector was 203 million tonnes of CO_2.

Our forest emissions in 2017 were equivalent to the annual carbon dioxide exhaust from 44 million passenger vehicles—almost double the Canadian passenger fleet.[116] They eclipsed BC's emissions from burning fossil fuels, and from industry and agriculture—64 million tonnes in 2017.[117] That doesn't mean, however, that the steady pulse of CO_2 discharged by the burning of fossil fuels is insignificant. Because carbon dioxide is a stable molecule, those annual emissions have accumulated in the atmosphere and contribute to the destabilization of our climate and the warming of our world. But forests, which could be our great ally in fighting climate change, are now so stressed that they are adding to the problem. You can see this in the following graphs on page 259.[118]

On the coast, the fastest, surest way to keep forests as an ally in the fight against climate change is to protect our old-growth trees from clear-cut logging. According to Jens, coastal forests can store over 3,670 tonnes of CO_2 per hectare, one of the highest rates on earth.[119] Carbon accumulates in soil and other plants as well as in trees. Around the world, old-growth forests may account for as much as 10 percent of the global uptake of CO_2.[120] As Jens explains, "Almost 70 per cent of the carbon stored in a tree is accumulated in the second half of its life."[121] Yet we have failed to preserve this important resource. Between 1990 and 2015, Vancouver Island's old-growth rainforests declined by 30 percent, three times faster than primary forest loss in tropical rainforests.[122] "Shockingly, about half of the carbon stored in these ecosystems is lost due to clear-cut logging from exposed soils and large amounts of wood left behind," Jens writes.[123]

The economic value that the old-growth rainforests provide by inhaling CO_2 is now recognized. In 2016, coastal First Nations signed an agreement with the Province of British Columbia that allows them to sell carbon credits for ending destructive logging practices in the rainforests. Instead of earning money by selling timber, they earn it for storing carbon. So far, the decreased rate of logging has resulted in the reduction of 2.2 million tonnes of CO_2 emissions annually.[124] This innovative arrangement could provide a model for further agreements to protect our forests and their capacity to absorb CO_2.

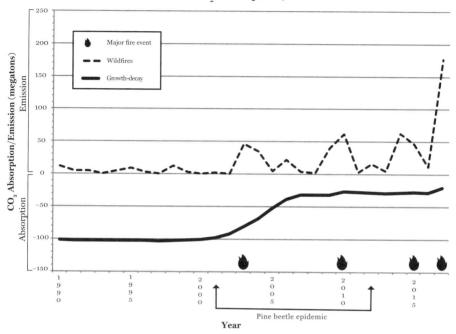

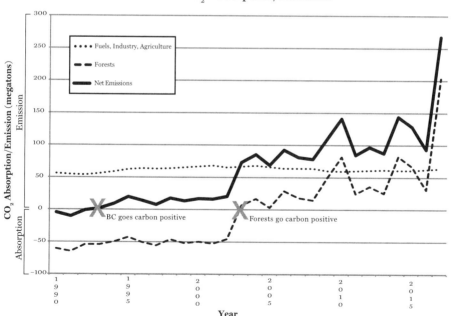

In less than thirty years, BC went from being a carbon sink to a carbon emitter.

Furthermore, the old-growth forests are singularly resistant to fire. You can easily see how those shady woods would keep the ground cool and moist. But there is more going on than meets the eye, as Peter Wohlleben tells us in *The Hidden Life of Trees*:

> Coniferous forests in the Northern Hemisphere influence climate and manage water in other ways, too. Conifers give off terpenes, substances originally intended as a defense against illness and pests. When these molecules get into the air, moisture condenses on them, creating clouds that are twice as thick as the clouds over non-forested areas. The possibility of rain increases, and in addition, about 5 percent of the sunlight is reflected away from the ground. Temperatures fall.[125]

As well as protecting these magnificent stands of ancient trees, we can accelerate the regrowth of forests decimated by insects or fires. Over the next four years, a Forest Carbon Initiative supported by BC and the federal government will put $290 million toward this goal, according to Werner Kurz at Natural Resources Canada. Timber companies are obliged to replant after they have harvested, but burned sites are usually left to regenerate on their own. Werner explained that if they are replanted, they become carbon sinks more quickly. We can also fertilize forests to speed up growth, although this has to be done carefully, as not all stands will respond. "Assisted migration" is another strategy for creating forests more resilient to global warming: "You take a tree species," Werner explained, "and move it northward by twenty to fifty kilometres, or maybe upslope by a few hundred metres. It is already accustomed to the new climate it is likely to experience in its lifetime."

"Hedging your bets" by planting a mix of conifers and deciduous trees is a sound principle for creating resilience in a situation that is inherently uncertain. It is also a sound principle in relation to carbon capture. A recent international study involving sixty researchers and 150,000 trees compared plots planted with between one and sixteen species. After eight years, the scientists found that the least diverse plots absorbed forty tonnes of CO_2 per hectare, while the most diverse plots took in more than twice that much—106 tonnes of CO_2.[126] Heterogeneity should be our by-word.

Suzanne Simard, a professor of forestry at UBC, discovered an underground mycelial network that links the roots of paper birches and Douglas firs. This may help explain why carbon absorption increases when a stand of trees is more diverse. Simard writes:

> Douglas firs were receiving more photosynthetic carbon from paper birches than they were transmitting, especially when the firs were in the shade of their leafy neighbours. This helped explain the synergy of the pair's relationship. The birches, it turned out, were spurring the growth of the firs, like carers in human social networks. Looking further, we discovered that the exchange between the two tree species was dynamic: each took different turns as "mother," depending on the season. And so they forged their duality into a oneness, making a forest.[127]

IN 1896, SVANTE ARRHENIUS, A Swedish physicist sometimes called "the father of climate change," estimated how much we could expect the surface temperature of our planet to rise if the amount of CO_2 in our atmosphere increased.[128] Were the draftees of the early Swedish Forestry Act aware of this? Had they heard of climate change? Did they consider that preserving forests might mitigate its effects? I was interested to note that Arrhenius received the Nobel Prize for Chemistry in 1903, the same year the Forestry Act was passed in Sweden.

We have taken our forests for granted. They were vast and must have seemed indestructible to the early settlers whose fires H.R. MacMillan decried. In two generations, he said we lost "700 billion board feet [1.65 billion cubic metres] of merchantable timber," about twenty-five times today's annual allowable cut. Suzanne Simard and Peter Wohlleben made me realize that forests are subtle places with many secrets. And now we stand at a crossroads, where we may lose them, just as we are beginning to understand what they are. To echo our lumber baron once again, "There is no record in history of such a loss." The forests are essential in our fight against global warming. They inhale our exhalations of CO_2, sequester them and grow to the sky. It behooves us to remember that we inhale *their* exhalations. I saw one study showing that thirty trees are needed to offset the annual oxygen consumption of an average adult.[129] The word "conspire" is from the Latin *conspirare*, *con-* "together with" and *spirare*, "breathe." Let us conspire with the forests, then. Let us preserve and protect them for our mutual benefit.

EPILOGUE

MAY 2019

IT IS NOW ALMOST TWO years since the fire went through. There have been changes. The new owners of Chris's property at the end of the lake are not running it as a resort, which means we can't launch our boat there to access our cabin from the lake. We had to build a driveway to the dirt road behind our property. Now we turn off the paved Sheridan Lake Road West to the unpaved Sheridan West Forest Service Road. At first, we see green on both sides of it, but past the 2.5 kilometre mark we begin to see burn, and after the 3.5 kilometre point, an expanse of burn. We can see Number Two Lake, which was always hidden before, behind a screen of trees. The fire has opened vistas.

The wrecked fences have been replaced, and the stretch where the fire was so hot and heavy that even the stumps were consumed is now a bright green meadow. Dandelions and small purple flowers festoon our way in. The firefighters had logged to create guards and piled the timbers on the side of the road. These have been taken away since we last drove through and I am pleased that someone has found a use for them. But there are still slash piles of unusable brush, branches and skinny trees. At seven kilometres, the black trunks and red needles give way to green again. We arrive at a patch that was attacked by the pine beetle and then logged. We had received a notice about that, telling us it was going to be a selective harvest. Only two trees survived the cut, but they are magnificent—two stately Douglas firs on either side of the road as we near our own property. I am grateful they were spared. To me they seem like sentinels, watchers, presiding over the forest. After the cut, the patch was replanted—all pine, now about two metres high—very close together. In Sweden, I think, someone would be thinning it.

After 8.6 kilometres, we turn onto our own driveway. There we see Jeff Sanson's handiwork. We hired him to manage a portion of the blowdown. He cut up the best logs and piled them, to be used as firewood. Then he amassed the brush and scrap into six enormous stacks and, in March, when snow was still on the ground, set fire to most of it. One small pile remains. Already the burned ground is filling in with grass. The deer will like it.

We get to our parking spot. We didn't put it next to our cabin. It's about seventy metres away, so you can't see our car from the cottage. This allows us to indulge in the illusion that our place is a refuge apart from the world. But I know there are no safe havens in reality. We can't escape the trends that affect everyone else. Nothing is apart. John Donne famously wrote, "No man is an island entire of itself; every man is a piece of the continent, a part of the main." These days, no island is an island, either.

We will treasure and try to protect what we have. But I know that won't be easy.

SELECT BIBLIOGRAPHY

2017 Provincial Inventory, "1990–2017 Greenhouse Gas Emissions Summary for British Columbia." Climate Action Secretariat. See: https://www2.gov.bc.ca/gov/content/environment/climate-change/data/provincial-inventory.

2019 National Inventory Report 1990–2017: Greenhouse Gas Sources and Sinks in Canada. Canada's Submission to the United Nations Framework Convention on Climate Change: Executive Summary. See: http://publications.gc.ca/collections/collection_2019/eccc/En81-4-1-2017-eng.pdf.

Abbott, George, and Chapman, Maureen. *Addressing the New Normal: 21st Century Disaster Management in British Columbia.* April 30, 2019. See: https://www2.gov.bc.ca/assets/gov/public-safety-and-emergency-services/emergency-preparedness-response-recovery/embc/bc-flood-and-wildfire-review-addressing-the-new-normal-21st-century-disaster-management-in-bc-web.pdf.

Anderson, Nancy Marguerite. "1843 Brigade Trail: Green Lake to Horse Lake," March 6, 2017. See: http://nancymarguerite-anderson.com/horse-lake/.

Briggs, Trevor. "Remembering 2017." BC Wildfire Service. See: https://www2.gov.bc.ca/gov/content/safety/wildfire-status/about-bcws/wildfire-history/remembering-2017.

Filmon, Gary. *Firestorm 2003: Provincial Review.* See: https://www2.gov.bc.ca/assets/gov/public-safety-and-emergency-services/wildfire-status/governance/bcws_firestorm report_2003.pdf.

"Forest Fire." *The Canadian Encyclopedia.* See "Fire Behaviour": https://www.thecanadianencyclopedia.ca/en/article/forest-fire#FireBehaviour.

"Forests and Forestry in Sweden." Royal Swedish Academy of Agriculture and Forestry. See: https://www.skogsstyrelsen.se/globalassets/in-english/forests-and-forestry-in-sweden_2015.pdf.

Gulli, Cathy. "What You Need to Know About the Science of Forest Fires." *Maclean's,* May 9, 2016. See: https://www.macleans.ca/society/science/qa-what-you-need-to-know-about-the-science-of-forest-fires/.

Keller, Keith. *Wildfire Wars.* (Madeira Park, BC: Harbour Publishing, 2002.)

Kirchmeier-Young, M.C., et al. "Attribution of the Influence of Human-Induced Climate Change on an Extreme Fire Season." *Advancing Earth and Space Science*, December 13, 2018. See: https://agupubs.onlinelibrary.wiley.com/doi/full/10.1029/2018EF001050.

Lewis, Brent. *brents desk*, YouTube. See: https://www.youtube.com/channel/UCJSEjCM_satXortC-d48P4w.

Official Report of Debates of the Legislative Assembly, March 3, 2016, Afternoon Sitting, Vol. 34, No. 3. See: https://www.leg.bc.ca/content/Hansard/40th5th/20160303pm-Hansard-v34n3.htm#.

Official Report of Debates of the Legislative Assembly, September 21, 2017, Morning Sitting, No. 23. See: https://www.leg.bc.ca/documents-data/debate-transcripts/41st-parliament/2nd-session/20170921am-Hansard-n23.

Parminter, John. "Human Influence on Landscape Pattern in the Pacific Region: Impacts of Burning by First Nations and Early European Settlers." Presented at the Landscape Ecology Symposium, 76th Annual Meeting of the Pacific Division, American Association for the Advancement of Science, Vancouver, BC, June 20, 1995. See: https://www.for.gov.bc.ca/hfd/pubs/Docs/scv/scv244.pdf.

"Wildfire Season Summary." BC Wildfire Service. See: https://www2.gov.bc.ca/gov/content/safety/wildfire-status/about-bcws/wildfire-history/wildfire-season-summary.

Winkleman, Max. "Fires Used to be Much More Common According to UBC Research," *The Williams Lake Tribune*, July 27, 2017. See: https://www.wltribune.com/news/fires-used-to-be-much-more-common-according-to-ubc-research/.

Wohlleben, Peter. *The Hidden Life of Trees*. (Vancouver/Berkeley: Greystone Books, 2015.)

ACKNOWLEDGEMENTS

MANY PEOPLE HELPED ME TO write this book. They took time out of their busy lives to tell me about events that were traumatic and troubling. I would like to thank Josh White for his descriptions of what happened at Ashcroft, and John Ranta and Tom Moe for their accounts of the fires in Cache Creek. Ryan Day increased my understanding of how the Bonaparte First Nation responded to the emergency, and Roger Hollander, Svend Nielsen, Val Severin and Mitch Campsall were invaluable for their knowledge about how 100 Mile House was affected. Heather Gorrell described the fires along the back roads around Williams Lake; Andra Holzapfel and Mark Taylor explained what was transpiring on the waterways in Bowron Lake Provincial Park. Max Forester, Jeremy Sieb and Craig Wilson gave me the perspective of smokejumpers near Quesnel, and Lana Shields revealed how over three hundred horses were rescued in Williams Lake. Brian McNaughton, Walt Cobb, Michel Legault, Dave Attfield, John Brewer, Jeremy Vogt, Gord Chipman, Samantha Smolen and Lee Todd provided insights into the saving of Williams Lake, and Pam Jim had an interesting story about the evacuation of that town. Blake Chipman, Ed and Noreen McDonald, Bryan and Raylene Poffenroth, Francis Laceese, and Kurt, Brenda and Chris Van Ember were great sources of information about the fires at Riske Creek. Lorraine and Dean Miller, and their two grandsons, Justin and Jordan Grier, explained how their family successfully defended the Chilco Ranch. Through Ryan Lake, Morgan and Kim Bosche, Jinwoo Kim, and Kenny and Teri-Lyn Dougherty, I was able to visualize events in Clinton. Ray Paulokangas, Alicia Polanski, Lorne Smith and Cheryl Merriman told me about their heartbreaking losses at Pressy Lake. Miguel and Krista Vieira, as well as their daughters, Makayla and Hannah,

ACKNOWLEDGEMENTS

gave me their unique view of how the fires affected 70 Mile House. Brad and Gail Potter presented the story of the inferno near Green Lake; Chris Brown, Carson Doward, Ryan Gahan and Gus Horn related how Sheridan Lake and the area around it was hit. Glen Burgess provided the viewpoint of an incident commander, and Brent Lewis described how, for a while, he came to be "the voice of the Cariboo." I would also like to thank Ray Travers, Robert Gray, Jens Wieting and Werner Kurz for their assistance with my final chapter, which looks at the lessons we can learn from the fires.

I'd also like to thank the photographers who generously shared their images with me: Bob Grant, Brad Pierro, Bailey Fuller, Wanda Shep, Jesaja Class, Steven Siebert, Shawn Cropley, Linda Botterill, Eric Depenau, Joanne Macaluso, Wolfgang Viertel, Bryan Johns and Kevin Haggkvist. Raylene Poffenroth, Krista Vieira, Barb Woodburn, Andra Holzapfel and Samantha Smolen not only consented to be interviewed but also gave me photos.

My husband, Gordon Cornwall, was the first person to read the manuscript and offered many valuable suggestions. He also took photographs of the people I interviewed and contributed pictures to the album in the centre of the book.

Finally, I'd like to thank my editors, Pam Robertson and Lynne Melcombe, as well as Howard White and Anna Comfort O'Keeffe at Harbour Publishing. I greatly appreciate their advice and faith in this project.

ENDNOTES

1   "Wildfire Season Summary," BC Wildfire Service. See: https://www2.gov.bc.ca/gov/content/safety/wildfire-status/about-bcws/wildfire-history/wildfire-season-summary.
2   Don Glass, "Fire Beetles," *A Moment of Science,* February 26, 2008. See: https://indianapublicmedia.org/amomentofscience/fire-beetles/.
3   John Parminter, "Human Influence on Landscape Pattern in the Pacific Region: Impacts of Burning by First Nations and Early European Settlers," presented at the Landscape Ecology Symposium, 76th Annual Meeting of the Pacific Division, American Association for the Advancement of Science, Vancouver, BC, June 20, 1995, pp. 2–4. See: https://www.for.gov.bc.ca/hfd/pubs/Docs/scv/scv244.pdf.
4   Ibid. p. 6.
5   Ibid. p. 9.
6   "BC Forest Service Timeline," BC Forest Service Centenary website. See: http://www.bcfs100.ca/bscripts/timeline.asp.
7   Parminter, "Human Influence," p. 5.
8   Gary Filmon, *Firestorm 2003: Provincial Review*, p. 32. See: https://www2.gov.bc.ca/assets/gov/public-safety-and-emergency-services/wildfire-status/governance/bcws_firestormreport_2003.pdf.
9   Ibid., p. 20.
10 Ibid., p. 10.
11 Ibid., pp. 24–33.
12 "Fuel Management in the Wildland Urban Interface—Update," Forest Practices Board, May 2015. See: https://www.bcfpb.ca/wp-content/uploads/2016/04/SIR43-Fuel-Management-Update.pdf.
13 Official Report of Debates of the Legislative Assembly, March 3, 2016, Vol. 34, No. 3. See: https://www.leg.bc.ca/content/Hansard/40th-5th/20160303pm-Hansard-v34n3.htm#.

14 "Climate Summary for: Cariboo Region," Pacific Climate Impacts Consortium, University of Victoria. See: https://www.pacificclimate.org/sites/default/files/publications/Climate_Summary-Cariboo.pdf.

15 David M. Romps, Jacob T. Seeley, David Vollaro and John Molinari, "Projected Increase in Lightning Strikes in the United States due to Global Warming," *Science* 14, November 2014: Vol. 346, No. 6211, pp. 851–54. See: http://science.sciencemag.org/content/346/6211/851.

16 "History of Mountain Pine Beetle Infestation in B.C.," Ministry of Forests, Lands, & Natural Resource Operations. See: https://www2.gov.bc.ca/assets/gov/farming-natural-resources-and-industry/forestry/forest-health/bark-beetles/history_of_the_mountain_pine_beetle_infestation.pdf.

17 "Mountain Pine Beetle (Factsheet)," Natural Resources Canada. See: https://www.nrcan.gc.ca/forests/fire-insects-disturbances/top-insects/13397.

18 "Effects of Climate Change on the Impact of Spruce Budworm Infestations," *Frontline Express*, Canadian Forest Service–Great Lakes Forestry Centre, Bulletin 61, NRC. See: http://publications.gc.ca/collections/collection_2013/rncan-nrcan/Fo122-1-61-2012-eng.pdf.

19 R.I. Alfaro, E. Campbell and B.C. Hawkes, "Historical Frequency, Intensity and Extent of Mountain Pine Beetle Disturbance in British Columbia," Natural Resources Canada, 2010. Reproduced with the permission of the Department of Natural Resources, 2019. See: https://cfs.nrcan.gc.ca/publications?id=31405.

20 Max Winkleman, "Fires Used to be Much More Common According to UBC Research," *Williams Lake Tribune*, July 27, 2017. See: https://www.wltribune.com/news/fires-used-to-be-much-more-common-according-to-ubc-research/.

21 "Cariboo-Chilcotin Wildfires 2017," Tim Conrad, Butterfly Effect Communications, p. 9. See: https://cariboord.ca/uploads/wildfirereport/Wildfire_Consultation_Report.pdf.

22 W. Jean Roach, Suzanne W. Simard and Donald L. Sachs, "Evidence Against Planting Lodgepole Pine Monocultures in the Cedar-Hemlock Forests of Southeastern British Columbia," *Forestry*, Vol. 88, No. 3, July 2015, pp. 345–58. See: https://academic.oup.com/forestry/article/88/3/345/678299.

23 "District of 100 Mile House Evacuation Order." See: http://www.100milehouse.com/files/2814/9965/8731/Evacuation_ORDER.pdf.

ENDNOTES

24 "Wildfire Response," Wildfire Service. See: https://www2.gov.bc.ca/gov/content/safety/wildfire-status/about-bcws/wildfire-response.
25 "Climate-Data.org: Ashcroft Climate," See: https://en.climate-data.org/north-america/canada/british-columbia/ashcroft-566778/#climate-table.
26 Government of Canada > Past Weather and Climate > Historical Data > Ashcroft. See: http://climate.weather.gc.ca/historical_data/search_historic_data_stations_e.html?searchType=stnName&timeframe=1&txtStationName=ashcroft&searchMethod=contains&StartYear=1840&EndYear=2019&optLimit=specDate&Year=2017&Month=7&Day=7&selRowPerPage=25.
27 According to Mac Gregory, executive director of the Volunteer Firefighters Association of BC, BC has four hundred fire departments staffed by volunteers, and a total of thirty thousand volunteer firefighters. Some are paid by the municipality in which they live; others are not.
28 "BC Geographical Names." See: http://apps.gov.bc.ca/pub/bcgnws/names/9342.html.
29 Ethan Siegel, "The Terrifying Physics of How Wildfires Spread So Fast," *Forbes*, September 6, 2017. See: https://www.forbes.com/sites/startswithabang/2017/09/06/the-terrifying-physics-of-how-wildfires-spread-so-fast/#2904cb037791.
30 News 1130, "Two First Nations Defy Evacuation Orders," July 12, 2017. See: https://www.citynews1130.com/2017/07/12/two-first-nations-defy-evacuation-orders/.
31 Keith Keller, *Wildfire Wars* (Madeira Park, BC: Harbour Publishing, 2002).
32 Crown-Indigenous Relations and Northern Affairs Canada. See: https://www.canada.ca/en/crown-indigenous-relations-northern-affairs.html.
33 Jon Azpiri, "B.C. Wildfire Status Saturday: Concerns Mount as Elephant Hill Wildfire Grows to 110,000 Hectares," Global TV, August 5, 2017. See: https://globalnews.ca/news/3650973/b-c-wildfire-saturday-concerns-mount-as-elephant-hill-wildfire-grows-to-110000-hectares/.
34 One hundred knots per hour.
35 The smokejumpers gave me imperial numbers, but for those who think metric, here is the conversion: 50 yards is 45 metres, 1,500 yards is 1,350 metres, 100 yards is 90 metres, 3 mph is 5 kph and 45 mph is 70 kph.
36 A medication to treat Parkinson's in horses.
37 Trevor Briggs, "Remembering 2017," BC Wildfire Service. See: https://www2.gov.bc.ca/gov/content/safety/wildfire-status/about-bcws/wildfire-history/remembering-2017.

38 Monica Lamb-Yorski, "Cariboo Fire Centre Wins Building Award," *Williams Lake Tribune*, March 15, 2018. See: https://www.wltribune.com/news/cariboo-fire-centre-wins-building-award/.

39 Autumn Macdonald, "Cariboo-Chilcotin Fire Centre Opens New Building," *Quesnel Cariboo Observer*, April 26, 2017. See: https://www.quesnelobserver.com/news/cariboo-chilcotin-fire-centre-opens-new-building/.

40 Ashley Wadhwani, "Williams Lake, 150 Mile House Under Evacuation," *Williams Lake Tribune*, July 15, 2017. See: https://www.wltribune.com/news/breaking-williams-lake-is-under-evacuation/?fbclid=IwAR1vwiWv_gNgRYIzwU62s44yk8Je1wve_drfEu3sjeiT5keVw-tI5PB9X-w.

41 For the metric-centric, here's the conversion: 100 yards is 90 metres, 120 feet is 35 metres and 30–40 feet is up to 12 metres.

42 George Abbott and Maureen Chapman, *Addressing the New Normal: 21st Century Disaster Management in British Columbia*, April 30, 2019, p. 85. See: https://www2.gov.bc.ca/assets/gov/public-safety-and-emergency-services/emergency-preparedness-response-recovery/embc/bc-flood-and-wildfire-review-addressing-the-new-normal-21st-century-disaster-management-in-bc-web.pdf.

43 Veera Bonner, Irene E Bliss and Hazel Henry Litterick, *Chilcotin: Preserving Pioneer Memories* (Surrey, BC: Heritage House Publishing, 1995), p. 149.

44 DECISION NO. 2005-FOR-016(a) In the matter of an appeal under section 82 of the Forest and Range Practices Act, S.B.C. 2002, c. 69, Forest Appeals Commission, p. 2. See: http://www.fac.gov.bc.ca/forestAndRange/2005for016a.pdf.

45 Charlie Smith, "B.C. Wildfires Continue to Rage out of Control," *The Georgia Straight*, July 9, 2017. See: https://www.straight.com/news/934316/bc-wildfires-continue-rage-out-control.

46 Deanna Kristensen, "Western Home—Chilco Ranch," *Western Horse Review*, May 12, 2015. See: http://www.westernhorsereview.com/cowboy-culture/embracing-the-chilcotin/.

47 "Forest Fire," *The Canadian Encyclopedia*. See "Fire Behaviour": https://www.thecanadianencyclopedia.ca/en/article/forest-fire#FireBehaviour.

48 On average, a BC household uses over ten thousand kilowatt hours of electricity per year, enough to power fifteen homes for a year with the energy discharged by one frontal metre of a high-intensity fire in an hour. "What is the average power use for a residential customer?" BC

Hydro Power Smart. See: https://www.bchydro.com/search.html?q=What+is+the+average+power+usage+for+a+residential+customer%3F&qid=1429&ir_type=3.

49 Abbott and Chapman, *Addressing the New Normal*, p. 99.

50 Barry Sale, "Haphazard History: Rich History Behind the 'Evans Place' Near Williams Lake," *Williams Lake Tribune*, January 7, 2019. See: https://www.wltribune.com/community/haphazard-history-rich-history-behind-the-evans-place-near-williams-lake/.

51 Ida Makaro, "Hundreds Gather to Celebrate Historic Maiden Creek Ranch," *Ashcroft-Cache Creek Journal*, August 18, 2012. See: https://www.ashcroftcachecreekjournal.com/community/hundreds-gather-to-celebrate-historic-maiden-creek-ranch/. And Barbara Roden, "Maiden Creek Ranch, Gold Country Geo Tourism Program." See: http://www.exploregoldcountry.com/pdf/caches/Maiden%20creek%20Ranch.pdf.

52 Patrick Johnston, "Controlled Burn North of Cache Creek Goes Wrong, Threatening Homes and Worrying Residents," *Vancouver Sun*, August 5, 2017. See: http://vancouversun.com/news/local-news/controlled-burn-goes-wrong-threatening-homes-and-worrying-residents.

53 *2017 Wildfires—Opportunities to Minimize Rural Impacts*, September 30, 2018. See: https://www2.gov.bc.ca/assets/gov/public-safety-and-emergency-services/emergency-preparedness-response-recovery/embc/2017-wildfires-opportunities-to-minimize-rural-impacts.pdf.

54 Mike Hager, "B.C. Officials Probe 'Back Burn' Started by Firefighters that Ranchers Say Destroyed Land," *The Globe and Mail*, August 9, 2017. See: https://www.theglobeandmail.com/news/british-columbia/bc-officials-probe-back-burn-started-by-firefighters-that-ranchers-say-destroyed-land/article35934000/.

55 Official Report of Debates (Hansard), September 21, 2017, Morning Sitting, No. 23. See: https://www.leg.bc.ca/documents-data/debate-transcripts/41st-parliament/2nd-session/20170921am-Hansard-n23.

56 "Response Package, pss-2017-73509," p. 9. See: http://docs.openinfo.gov.bc.ca/Response_Package_pss-2017-73509.pdf.

57 Ibid., pp.19–151.

58 Ibid., p. 8.

59 363664009-Pressy-Lake-foi-Documents, Scribd, p. 88. See: https://www.scribd.com/document/363664009/Pressy-Lake-foi-Documents.

60 Ibid, p. 21.

ENDNOTES

61 Ibid, p. 74.
62 "Wildfire Rank," BC Wildfire Service. See: https://www2.gov.bc.ca/gov/content/safety/wildfire-status/about-bcws/wildfire-response/fire-characteristics/rank.
63 Pressy Lake FOI documents, p. 42.
64 Abbott and Chapman, *Addressing the New Normal*, p. 11.
65 Ann Hui, "Some B.C. Food Stores Empty, Others Overstocked Amid Wildfires," *The Globe and Mail*, July 13, 2017. See: https://www.theglobeandmail.com/news/british-columbia/some-bc-food-stores-empty-others-overstocked-amid-wildfires/article35678430/.
66 *Green Lake and Area Official Community Plan*, amended July 17, 2014, p. 21. See: https://tnrd.civicweb.net/document/69444.
67 Max Winkleman, "July 22: Elephant Hill Fire Has Not Grown Outside of its Perimeter Today," *Ashcroft-Cache Creek Journal*, July 22, 2017. See: https://www.ashcroftcachecreekjournal.com/news/july-22-elephant-hill-fire-has-not-grown-outside-of-its-perimeter-today/.
68 Pressy Lake FOI documents, p. 13.
69 Ibid., p. 92.
70 Ibid., p. 17.
71 "Bombardier Canadair 415 Superscooper," *Aerospace Technology*. See: https://www.aerospace-technology.com/projects/bombardier_415/.
72 "Wildfire Aviation," BC Wildfire Service. See: https://www2.gov.bc.ca/gov/content/safety/wildfire-status/about-bcws/wildfire-response/response/aviation.
73 "Niagara Falls Facts," Niagara Imax Theatre. See: https://imaxniagara.com/niagara-falls-facts/.
74 Nancy Marguerite Anderson, "1843 Brigade Trail: Green Lake to Horse Lake," March 6, 2017. See: http://nancymargueriteanderson.com/horse-lake/.
75 Elephant Hill (K20637) Public Information Map, August 17, 2017. See: http://bcfireinfo.for.gov.bc.ca/ftp/!Project/Wildfire-News/8172017~35407_17%20K20637%20MAP%20INFO%20Public%20500K%208.5x11P%2020170817.pdf.
76 Elephant Hill (K20637) Public Information Map, August 25, 2017. See: http://bcfireinfo.for.gov.bc.ca/ftp/!Project/Wildfire-News/8252017~100013_17%20K20637%20MAP%20INFO%20Public%20500K%208.5x11P%2020170825.pdf.

ENDNOTES

77 Trevor Pugh, "Water Bombers on Sheridan Lake," August 31, 2017. See: https://www.youtube.com/watch?v=QsJagLytnlE.

78 Elephant Hill (K20637) Public Information Map, August 31, 2017. See: http://bcfireinfo.for.gov.bc.ca/ftp/!Project/Wildfire-News/922017~124641_17%20K20637%20MAP%20INFO%20Public%20500K%208.5x11P%2020170902.pdf.

79 *brents desk*, YouTube. See: https://www.youtube.com/channel/UCJSEjCM_satXortC-d48P4w.

80 "Cathy Gulli. "What You Need to Know about the Science of Forest Fires," *Maclean's*, May 9, 2016. See: https://www.macleans.ca/society/science/qa-what-you-need-to-know-about-the-science-of-forest-fires/.

81 "Data Sources and Methods: Fire Monitoring, Mapping, and Modeling (Fire M3)," Natural Resources Canada. See: http://cwfis.cfs.nrcan.gc.ca/background/dsm/fm3.

82 Don Schwartz, South Green Lake Community Group, Facebook, September 2, 2017. See: https://www.facebook.com/groups/969178076498853/permalink/1460089257407730/.

83 Elephant Hill Public Information Map, September 4, 2017. See: http://bcfireinfo.for.gov.bc.ca/ftp/!Project/WildfireNews/942017~24057_17%20K20637%20MAP%20INFO%20Public%20500K%208.5x11P%2020170904.pdf.

84 "Inter-Agency Operational Procedures and Reimbursement Rates: Between the Office of the Fire Commissioner, The Fire Chiefs Association of BC, BC Wildfire Service," revised July 2017. See: https://www2.gov.bc.ca/assets/gov/public-safety-and-emergency-services/emergency-preparedness-response-recovery/embc/policies/inter-agency_working_group_report_reimbursement_rates_2018.pdf.

85 A partial list includes Four Paws International, the Canadian Disaster Animal Response Team (CDART), the United Way, Kamloops Food Bank, Thompson Rivers University, the Canadian Red Cross, ASK Wellness Society, 4H Clubs, the Salvation Army, BC Lottery Corporation, the Disaster Psychosocial Services Program, the Ministry of Social Development and Social Innovation, and St. John's Ambulance Services.

86 "Administration Board Report," Thompson-Nicola Regional District, October 31, 2017. Obtained via email from Agnese Saat, Grants Co-ordinator for the TNRD.

87 Abbott and Chapman, *Addressing the New Normal*.

88 Dominique Boucher et al., "Current and Projected Cumulative Impacts of Fire, Drought, and Insects on Timber Volumes Across Canada," *Ecological Applications*, April 12, 2018. See: https://esajournals.onlinelibrary.wiley.com/doi/abs/10.1002/eap.1724.

89 Statistics are all from "Wildfire Season Summary," BC Wildfire Service. See: https://www2.gov.bc.ca/gov/content/safety/wildfire-status/about-bcws/wildfire-history/wildfire-season-summary.

90 Robert Gray, "Total Fire Cost: The Rationale and Value of Increased Investment in Prevention," *Climate Extremes in BC: A New Environmental Reality*, November 16, 2017. See: https://www.emaofbc.com/wp-content/uploads/2017/11/Total-Fire-Cost_rwg.pdf.

91 Nelson Bennett, "Fires Claim a Year's Worth of Timber in the Province," *Business In Vancouver*, September 26, 2017. See: https://biv.com/article/2017/09/fires-claim-years-worth-timber-province.

92 Derrick Penner, "B.C. Fights Trade Headwinds on Forestry Sales Mission to Asia," *Vancouver Sun*, December 4, 2018.

93 Gray, "Total Fire Cost."

94 Barb Glen, "B.C. Fires Leave Livelihoods of Farmers, Ranchers in Ashes," *The Western Producer*, September 21, 2017. See: https://www.producer.com/2017/09/b-c-fires-leave-livelihoods-of-farmers-ranchers-in-ashes/.

95 Gray, "Total Fire Cost."

96 Linda Givetash, "Hospital Visits Jump in B.C. as Smoky Air from Wildfires Persists," *Vancouver Sun*, August 5, 2017.

97 *Government's Action Plan: Responding to Wildfire and Flood Risks*, October 31, 2018. See: https://www2.gov.bc.ca/assets/gov/public-safety-and-emergency-services/emergency-preparedness-response-recovery/embc/action_plan.pdf.

98 M. C. Kirchmeier-Young et al., "Attribution of the Influence of Human-Induced Climate Change on an Extreme Fire Season," *Advancing Earth and Space Science*, December 13, 2018. See: https://agupubs.onlinelibrary.wiley.com/doi/full/10.1029/2018EF001050.

99 Ray's comment is confirmed in the following reports: For British Columbia, B.C. Ministry of Forests, Mines and Lands, *The State of British Columbia's Forests*, 3rd ed., 2010. Forest Practices and Investment Branch, Victoria, B.C., p. 18. See: https://www2.gov.bc.ca/assets/gov/environment/research-monitoring-and-reporting/reporting/envreportbc/archived-reports/sof_2010.pdf. And for Sweden, "Brief Facts 1: What is Swedish

Forestry?" *Sveaskog*. See: https://www.sveaskog.se/en/forestry-the-swedish-way/short-facts/brief-facts-1/.

100 Catherine Edwards, "Sweden's Green Soul: Why Forests are Vital to the Swedish Culture and Economy," *The Local*, July 24, 2019. See: https://www.thelocal.se/20180724/sweden-forest-wildfire-economy-trees-culture-lifestyle.

101 Daniel Pierce, "25 Years After the War in the Woods: Why B.C.'s Forests are Still in Crisis," *The Narwhal*, May 14, 2018. See: https://thenarwhal.ca/25-years-after-clayoquot-sound-blockades-the-war-in-the-woods-never-ended-and-its-heating-back-up/.

102 "Brief Facts 1."

103 "Facts & Figures: Sweden's Forest Industry in Brief." See: https://www.forestindustries.se/forest-industry/facts-and-figures/.

104 "Land & Forests: Trends in Timber Harvest in B.C.," *Environmental Reporting BC*. See: http://www.env.gov.bc.ca/soe/indicators/land/timber-harvest.html.

105 Bob Williams, *Restoring Forestry in BC*, Centre for Policy Alternatives, January 22, 2018, p. 6. See: https://www.policyalternatives.ca/sites/default/files/uploads/publications/BC%20office/2018/01/CCPA-BC_RestoringForestry_web.pdf.

106 Ibid. p. 6.

107 Torbjörn Johnsen, "Forest Fires in Sweden—Huge Areas Burned in 2018," *Forestry.com*, August 6, 2018. See: https://www.forestry.com/editorial/forest-fires-sweden/.

108 *Forests and Forestry in Sweden*, Royal Swedish Academy of Agriculture and Forestry. See: https://www.skogsstyrelsen.se/globalassets/in-english/forests-and-forestry-in-sweden_2015.pdf.

109 Karin Ericsson and Sven Werner, "The Introduction and Expansion of Biomass Use in Swedish District Heating Systems," *Biomass and Bioenergy*, Vol. 94, November 2016, pp. 57–65. See: https://www.sciencedirect.com/science/article/pii/S0961953416302793.

110 Tildy Bayar, "Sweden's Bioenergy Success Story," *Renewable Energy World*, Vol. 16, No. 1, March 13, 2013. See: https://www.renewableenergyworld.com/articles/print/volume-16/issue-1/bioenergy/swedens-bioenergy-success-story.html.

111 "British Columbia Wood Pellets: Sustainability Fact Sheet," Wood Pellet Association of Canada, p. 4. See: https://www.canadianbiomassmagazine.ca/images/bc-biomass.pdf.

112 "Fire on the Mountain: Rethinking Forest Management in the Sierra Nevada," Little Hoover Commission, Report 242, February 2018. See: https://lhc.ca.gov/sites/lhc.ca.gov/files/Reports/242/Report242.pdf.

113 S.G. Cummings, "Forest Type and Wildfire in the Alberta Boreal Mixedwood: What Do Fires Burn?" *Ecological Applications*, Vol. 11, No. 1, February 2001, pp. 97–110. See: http://stopthespraybc.com/wp-content/uploads/2011/07/Cumming.2001.EcolAppl.pdf.

114 2017 Provincial Inventory, "1990–2017 Greenhouse Gas Emissions Summary for British Columbia." Climate Action Secretariat. See: https://www2.gov.bc.ca/gov/content/environment/climate-change/data/provincial-inventory.

115 Ibid.

116 "Greenhouse Gas Emissions from a Typical Passenger Vehicle," United States Environmental Protection Agency. See: https://www.epa.gov/greenvehicles/greenhouse-gas-emissions-typical-passenger-vehicle. A typical passenger vehicle emits 4.6 tonnes of CO_2 a year; 203 million tonnes of CO_2 divided by 4.6 tonnes of CO_2 is 44 million. For the size of passenger fleet, see "Vehicle Registrations, by Type of Vehicle," Statistics Canada, https://www150.statcan.gc.ca/t1/tbl1/en/tv.action?pid=2310006701.

117 2017 Provincial Inventory.

118 Graphs: Gordon Cornwall.

119 Jens Wieting, "Preserve Old-Growth Forests to Keep Carbon Where It Belongs," *The Tyee*, September 12, 2017. See: https://thetyee.ca/Opinion/2017/09/12/Preserve-Forests-Carbon-Belongs/.

120 "Old Growth Forests Are Valuable Carbon Sinks," *ScienceDaily*, Oregon State University, September 14, 2008. See: https://www.sciencedaily.com/releases/2008/09/080910133934.htm.

121 Wieting, "Preserve Old-Growth Forests."

122 "Twenty-Five International Organizations Call for Urgent Action for Vancouver Island's Rainforest and Communities," Sierra Club BC, April 27, 2019. See: https://sierraclub.bc.ca/25-organizations-call-for-action-vancouver-islands-rainforest-communities/.

123 Wieting, "Preserve Old-Growth Forests."

124 Ibid.

125 Peter Wohlleben, *The Hidden Life of Trees* (Vancouver/Berkeley: Greystone Books, 2015), p. 107.

126 Adam Aton, "Diverse Forests Capture More Carbon," *E & E News, Scientific American*, October 5, 2018. See: https://www.scientificamerican.com/article/diverse-forests-capture-more-carbon/.

127 Suzanne Simard, "Notes from a Forest Scientist" (afterword), in Peter Wohlleben, *The Hidden Life of Trees*, p. 248.

128 Ian Sample, "The Father of Climate Change," *The Guardian*, June 30, 2005. See: https://www.theguardian.com/environment/2005/jun/30/climatechange.climatechangeenvironment2.

129 David J. Nowak, Robert Hoehn and Daniel E. Crane, "Oxygen Production by Urban Trees in the United States," *Arboriculture & Urban Forestry*, Vol. 33, No. 3, 2007, pp. 220–26. See: https://www.nrs.fs.fed.us/pubs/jrnl/2007/nrs_2007_nowak_001.pdf.

INDEX

Note: Page numbers in **bold** refer to photographs; page numbers in **bold** roman numerals refer to photographs in the insert.

16 Mile House, **VIII**, 35–43, 187, 189, 244
16 Mile House fire brigade, 43
20 Mile area, 41
70 Mile General Store, 197, 208, 211, 214–15
70 Mile House, **IX**, **X**, 48, 205–15, 254
70 Mile House Volunteer Fire Department, 211
99 Mile House, 209
100 Mile Free Press, 53, 55
100 Mile House, 9, 12–14, 44–59, 93, 105, 107, 111, 197, 208–11, 214, 229–30, 241, **242**, 247
100 Mile House Fire Department, 45, **46**, 47, 58
103 Mile House, 48–51
105 Mile House, 12, 47–48, 57
108 Mile House, **I**, 12, 47–49
108 Mile House Fire Department, 45, 49
140 Mile House, 6, 95
150 Mile House, 48, 64, 104

Abbotsford, 175
Abbott, George, 125, 161, 247

Aie, Heather, 247
Aie, Steve, 19–20
Alex Fraser Forest, 6
Alexis Creek, 49, 96, 155
Alfaro, René, 6
Alicia, 214
Alkali Lake, 88, 116, 120–21, 124, 162
Alkali Lake Indian Band. *See* Esk'etemc First Nation
Alkali Resource Management, 88, 116
Alkali Unit Crew, **119**, 121–22
Allen, Darby, 54
Anaheim, 49
Anderson, Alexander, 224
Arikado, Lyn, 246
Arrhenius, Svante, 261
Arrow Lakes, 233
Ashcroft, 9, 15–22, 28, 170, 186, 192, 195, 242, 246–47
Ashcroft First Nation, 16, 24, 29, 33
Ashcroft Journal, 217
Ashcroft Volunteer Fire Department, 16, 21
Atlin, 3
Attfield, Dave, **107**, 111, 113–14
Azpiri, Jon, 41

INDEX

Bains, Harry, 4
Balding, Dave, 246
Barkerville, 78, 185
Barrett, Dave, 254
Barriere, 14
BC Conservation Officer Service, 66, 200, 220, 243
BC Cowboy Hall of Fame, 31, 140
BC Fire Chiefs Conference, 54
BC Forest Service (formerly BC Forest Branch), 4, 62, 137
BC Freedom of Information and Protection Act (FOI), 203–6
BC Legislature, 4, 203
BC Ministry of Citizens' Services, 203
BC Ministry of Forests, Lands and Natural Resource Operations, 192
BC Office of the Fire Commissioner, 203, 245
BC Parks, 61, **62**, 63–64
BC RCMP Headquarters, 111
BC Wildfire Parattack Unit, 80, 83
BC Wildfire Service, 1, 4, 12–13, 17–19, 28–32, 45, 51, 58, 81, 88, 108, 118–30, 158, 161–70, 180–81, 187–92, 203–6, 219–21, 230–34, 239
Bear (stuffed toy), **212**–15
Bear River Mercantile, 71
Beaver Valley Feeds, 96
Becher, Fred, 157
Becher Ranch, **VIII**, 157–58, **159**, 160–66, 244
Becher's Dam, 164
Becher's Prairie, 164
Beddow Tree webcam, 237
Begbie, Matthew, 223–24
Begbie Trail, 223–24

Bella Coola, 155
Bennett, Ryland, 84
Big Creek, 130
Big Rock, the, 140
Big White, 171–72
Bigg, Dawn, 248
Blackwater Unit, 83–85, 103, 245
Blocks R Us, 124
Bonaparte Canyon, 20
Bonaparte First Nation, **II**, 24, 29, **30**, 31–33, 36–40
Bonaparte River, 169, 171, 217
Bonaparte Valley, 185
Bonner, Lynn, 141
Bosche, Kim, 174–76, **177**
Bosche, Morgyn, 174–76, **177**
Boston Flats Trailer Park, 20
Bougie, Brandon, 45
Boundary District, 3
Bowron Lake, 78
Bowron Lake Provincial Park, 1, 62–63, 71, **72**, 79, 249–50
Bowron Lake Provincial Park Campground, 79
Boyd Bay, 218
Bradner Farms, 19–20
Brassington, Ashley, 73, 75
brents desk, 235–36
Brewer, John, 111, **112**, 114
Brigade Trail, 112
Briggs, Trevor, 103–4
Brooke, Wesley, 6
Brown, Chris, 10, 14, **229**, 230–35, 238–40, 263
Bukowski, Charles, 244
Burgess, Glen, 1, 5, **25**, 110, 171, 188, 204–5, 219–20, 245

INDEX

Burns Lake, 115
Bush Fire Act (1874), 4

Cache Creek, 1, 12–17, 23–34, 180, 192
Cache Creek Volunteer Fire Department, 24, 32
Cade, Helene, 186
Campbell, Gordon, 4
Campsall, Mitch, **51**, 52–57
Canadian Armed Forces, x, 165, 213–14
Canadian Fire Service, 30
Canadian Press, 30
Canim Lake, 49, **249**
Canoe River, 3
Cariboo, 3, 5, 48, 104, 112, 178, 185, 209, 227, 250, 256
Cariboo Bethel Church, 113, 244
Cariboo Community Church, 113–14
Cariboo Emergency Notification System, 230, 249
Cariboo Fire Centre, 6, 61–62, 73, 81, 84, 87, 89, 103–4, 124–25, 149
Cariboo Gold Rush, 45, 78
Cariboo Highway, 14, 178
Cariboo Mountains Provincial Park, 61–62
Cariboo Regional District (CRD), 47–50, 98, 105, 160, 181
Cariboo River, 73–74
Cariboo Wagon Road, 24, 185, 224
Cariboo-Chilcotin, 117, 252
Cariboo-Chilcotin Plateau, 6
Carlson, Terri, 73, 77
Cassiar, 3
Cassidy, Clayton, 17
Castlegar, 233

CBC News, 98, 177
Century Ranch Award, 184
Chapman, Maureen, 125, 161, 247
Charge-Pacific Region Training Centre, 106
Charlie (cat), 13
Chase, 187
Chasm, 236, 254
Chasm Sawmill, 26, 186, 211–13
Chelsea, Glen, 118, **119**, 120–21
Cherry Creek Ranch, 249
Chetwynd Zone, 81
Chilco Lake, 4
Chilco Ranch, 129–30, **131**, 132, **133**, 134–35, 177, 244
Chilcotin, 103–5, 119, 135, 153, 256
Chilcotin Lodge, 146, **147**, 148, 155, 157, 162
Chilcotin River, 130, 135
Chilcotin Towing, VII, 139, **142**, 163
Chilliwack, 106, 150, 151, 252
Chipman, Blake, 120–21, 127, 142, 244
Chipman, Gord, 88, 116–27, 163
Christianson, Amy, 30
Class, Jesaja, 145
Class, Roland, VI
Class, Udette, VI
Classmates (band), 217
Clayoquot Sound, 253
Clearwater, 204
Clinkenmaillie, Catherine, **248**
Clinton, 13, 41, 168–93, 204–6, 210–11, 218, 247, **248**, 249
Clinton Shell Gas and Budget Foods, 175–81, **182**, 247
CN, 21
Cobb, Walt, 105–6, 126

INDEX

Columbia River, 3
Columbia Valley, **249**
Columeetza Secondary School, 109
Comox Fire Department, 200, 205
Conair Air Tractor 802 Fire Boss, **I**, **XIII**, **XIV**, 245
Constantia Resources, 38
Coquitlam, 203
Cornwall, Claudia, 9–14, 129, 137–38, 146, 157, 175, 184, 208, 217, 226–31, 249
Cornwall, Gordon, 9–14, 129–30, 137–38, 146, 157, 175, 184, 208, 217, 226–31, 242, 249
Cornwall, Talia, 11, 227
Cornwall, Tom, 10–11, 228
Cranbrook, **XIII**
Critical Mass Pop-Up Gallery, 241
Crown-Indigenous Relations and Northern Affairs Canada, 33
Cultus Lake, 249

Dan, Dave, 118, **119**
Daniels, Lori, 6
Dawson Creek, 162–63
Day, Chief Ryan, 29, **30**, 31–34
Deadman's River, 177
Deep Creek, 62, 65–66
Delta, 189–90
Delta Fire Rescue, **XI**
Department of Indigenous and Northern Affairs Canada, 33
Dirks, Wren, 84
Donaldson, Doug, 192
Donne, John, 264
Dorward, Carson, 234–35, 238
Dougherty, Chuck, 186

Dougherty, Edward, 184–85
Dougherty, Elizabeth, 185
Dougherty, Kenny, 184, **185**, 186–93
Dougherty, Ray, 41, 186
Dougherty, Teri-Lyn, 41, 184, **185**, 186–93
Dragon Mountain, 1, 81, 83
Drive BC, 13
Dunn Lake, 12
Dusty Rose Pub, 215
Dyck, Cory, 164, 244

Eden, Ron, 243
Eldorado Log Hauling, 122, **123**, 124–25, 163
Elephant Hill, **IV**, **XV**, 1, 19–24, **25**, 34, 41, 169–71, 181, 187, 204–5, 214, 217–19, 226–27, 233, 240, 242, 245
Emergency Operations Centres, 235
Environment and Climate Change Canada, 173, 253
Esk'etemc First Nation, 88–90

Facebook, 37–43, 93–99, 155, 179–84, 207, 214, 221–22, 232, 238–39, 250
Farwell Canyon, 157
Federation of BC Woodlot Associations, 105
Filmon, Gary, 4–5, 255
Fire Behaviour Forecast, 219
Fire Centres, 235
First Call protection unit, 204–5
First Nations, 3, 7, 24, 30, 33, 154, 258
First Nations Policing, 111
Fisher, Jane, 9–13, 201, 228–30, 242
Forest Carbon Initiative, 260
Forester, Max, 80–91

284

INDEX

Forestry Act (Sweden, 1903), 261
Fort McMurray, AB, 54
Fort St. James, 254
Fort St. John, 1, 80
Fountain Flat Trading Post, 182
Fox Mountain, 62–63, 94, 98, 105
Fraser Canyon, 13, 195
Fraser River, 1, 3, 81, 86, 106, 122, 242
Fraser River Bridge, 139
Fraser-Nicola electoral district, 203
Fuller, Brian, 148, **149**, 150, 152
Fuller, Evan, 157, 159, 164

G.F. Strong Rehabilitation Centre, 177–78
Gahan, Ryan, 233, **235**
Gall, Peter, 53
Gang Ranch, 161
Ghost Lake, 61, 63
Global TV, 41
Globe and Mail, 209
Gold Mountain Restaurant, 178
Golden, 246
Gorrell, Heather, 61, **62**, 63–69, 73–74, 244
Gray, Robert, 252–57
Green Lake, **XIV**, **XV**, 208–24, 233, **235**, 241. *See also* North Green Lake; South Green Lake
Green Lake Provincial Park, 218
Green Mountain, 103
Gregory, Mac, 272n27
Grier, Jordan, 130–32, **133**, 136–37
Grier, Justin, 130–32, **133**, 134–37, 177
Gustafsen Lake, 12, **46**, 51, 58, 93

Hanceville, **VI**, **VII**, 96

Hanceville-Riske Creek, 1, 128–67
Handsworth, 116
Harcourt, Mike, 254
Harnden, Shelly, 121
Harry, Brett, 118, **119**
Hart Ridge, 173, 176–77
Hart-Chief, 69
Hawkins Lake, 49
Hidden Life of Trees, The, 260
Highway 1, 26, 117, 205
Highway 5, 107–8
Highway 20, **VII**, 118–20, 130, 139, 148–50, 155, 165–66
Highway 24, 13–14, 54, 107, 112, 198, 211–13, 230, 234
Highway 97, **XI**, 9, 13, 26, 32–33, 38, 51–54, 67, 103, 106, 108, 113, 117, 181, 187–88, 213
Highway 97c, 117
Highway 99, 13, 117, 181, 205
Hihium Lake, 205
Hollander, Roger, 45, **46**, 54, 55–58, 247
Holzapfel, Andra, 1, 71, **72**, 73–79
Holzapfel, Jack, 71, **72**, 73–79
Holzapfel, Rick, 71, **72**, 73–79
Holzapfel, Tom, 71, **72**, 73–79
Horn, Gus, 241, **242**, 243–44
Horn, Helen, 242
Horse Council of BC, 102
Horse Fly, 248
Horse Lake, 217, 221, 224
Hudson's Bay Brigade Trails, 112, 224
Hudson's Bay Company, 3, 112, 224
Hui, Ann, 209
Husky Energy, 113
Hutchinson, Tom, 84

Hutchison Lake, 221
Hutchison Road, 220-23
Hytest Timber, 141, 163-64

Ilnicki, Tom, 158
Ilnicki Ranch, 158
Indian Act, 33
Indigenous Services Canada, 33
Indoor Rodeo, 94
Indy (horse), 152
Initial Attack Unit, 87
Interior (BC), 4, 108, 249
Interior Health Authority, 52
Interlakes area, 11
Interlakes Volunteer Fire Department, 216
Isaac Lake, 73-74

Jack Frost Lake, 231, 241
James, Malcolm, **249**
Jasper, Connie, 162, 167, 250
Jasper, Delmer, 140, 144
Jasper, Leland, 140, 165
Jasper, Lorraine, 140
Jasper, Mike, 139-40, 162, 167, 250
Jasper, Pat, 140, 165-66
Jasper, Wes, 140, 166
Jim, Jenny, 112
Jim, Kam, 108, 112
Jim, Pam, 108, 112-13, 247
Jim Lake, 204
Jim's Food Market, 108, 112, 247
Johnsen, Torbjörn, 255

K Division Tactical Unit (RCMP, Alberta), 107-8

Kamloops, 1, **2**, 4, 19, 29, 52, 107-8, 110, 113, 115, 169, 217, 222, 246
Kamloops Fire Centre, 1, 6
Kamloops Lake, 24
Kamloops Reception Centre, 246
Kane, Pat, 11, 227
Keener, George, XII, 125
Keller, Keith, 31
Kelowna, 254
Kim, Jinwoo, 175, 179-81, **182**, 183, 247
Kim, Sang, 180, 183
Kleena Kleene, 163
Kootenay River, 3
Kurosawa, Akira, 158
Kurz, Werner, 250, 260
Kuzyk, Brian, 43

Lac La Hache, 48, 60-69, 209
Laceese, Chief Francis, 1, **2**, 250
Lake, Ryan, 169-72, **173**, 174-75, 177-78, 181
Lamb, Brian, 49, 51
Lamb Creek, XIII, XIV
Lanezi Lake, 75-76
Langley, 197
Leduc Fire Services (AB), 201
Lee's Corner, VI, 129-30, 150
Legault, Michel, **106**, 107-15
Lewis, Brent, 235-37
Likely, 61, 63
Lillooet, 13, 45, 182
Lillooet Highway, 193
Little Fort, I, 9, 12-14, 54, 107-8, 112, 248
Little Green Lake, 230

INDEX

Little Hoover Commission (USA), 256
Little Horse Lake, 230
Lone Butte, 180, 209
Lone Butte Volunteer Fire Department, 45
Loon Lake, 14, 41, 187–88
Loon Lake Road, 187
Lorentz, Unni, 198–99
Lorentz, Wayne, 198–99
Loring, Hugh, 165, 167
Loring, Shelly, 165, 167
Lost Valley Road, 206
Lower Similkameen Indian Band, 111
Lyon, Dale, 18, 22
Lyon, Maggie, 18, 21

Mack, Jessica, 104
Mackenzie, 80, 254
MacMillan, H.R., 3–4, 261–62
Magnussen Road, 234
Maiden Creek, 41
Maiden Creek Ranch, 184, **185**, 186–91
Maple Ridge, 195, 254
Marshall's 150 Mile Store, 61
Mayer, Alena, 141, 143
McDonald, Ed, 139–41, **142**, 143–45, 163, 166
McDonald, Jason, 139–45
McDonald, Noreen, 139–41, **142**, 143–45, 166
McDonald Summit, 112
McGregor Mountains, 81
McLean, Don, 46–47
McLeary Lake, 73–74
McLeese Lake, 49, 103

McLeod, Sandra, 9, 12
McNaughton, Brian, 105, 108, 111, 115
McNulty, Old Man, 243
Meldrum Creek, 140
Merriman, Cheryl, 200
Midgley, Allen, 38
Milk Ranch, 140
Miller, Dean, 129–30, **131**, 132, **133**, 134–37, 244
Miller, Janet, 132
Miller, Lorraine, 129–30, 134, 136
Minnabarriet, Marie, 31
Minnabarriet, Percy, 31
Minton, Mike, 129
Mitchell, Leanna, 11–12, 29, 228
Mr. Twister (bull), 177
Moe, Tom, 24, 26–28
Mountain House, 66
Murray, Gordon, 255

Nath, Carl, 242–43
Nath, Sarah, 243
National Forestry Commission of Mexico (CONAFOR), XII
National Research Council Interactive Infrared Map, 237
Natural Resources Canada, 250, 260
Nazko, XIII, 103
Nelson, 171
Nemiah Valley, 145
News 1130, 30
Nielsen, Svend, 46, **47**, 48–59, 107–8, 111, 251
Noah (Bible), 67
North Bonaparte Road, 213
North Green Lake, 214

North Vancouver, 235, 237
North West Company, 3
Number Two Lake, 237, 263
Nyman, Greg, 188

Old Cariboo Road, 36
Old School, 121, 148, **149**, 155
Old Soda Creek Road, 99
Otter Lake, 1, 81
Owens, Rick, 204

Pacific Fire Centre, 250
Paradise Bay Resort, **229**, 230–31, 238–39
Parallel Wood Products, 93
Parminter, John, 3
Paulokangas, Anita, 195
Paulokangas, Annikki, 195–97
Paulokangas, Niilo, 195, **196**, 206
Paulokangas, Ray, 195, **196**, 197–99, 202, 206–7, 214
Paulokangas, Seija, 195
Pauls, Jeremy, 73, 75
Pavilion, 206
Pavilion Road, 181
Pederson, Heather, 203–4
Pemberton, 13
Perry Ranch, 37
Peterson Contracting, 163
Piesse, Logan, 100
Pinchbeck, Bridgette, 125
Pioneer Log Homes of BC, 216
Pioneer Logging, 64
Plateau Complex, 1–2
Platt, Nancy, 228–30, 239
Poffenroth, Bryan, 157–58, **159**, 160–67, 244, 250

Poffenroth, Raylene, 1, 157–60, **161**, 162–67, 244, 250
Polanski, Alicia, 198–203, 206, 236
Polanski, Tim, 198–99, 201
Port Coquitlam, 236
Port Moody, 111, 210
Potter, Brad, 216–18, **219**, 220–24, 231
Potter, Gail, 216–17, 221–23, 231
Pressy Lake, 194–211, 220, 236, 241
Prince George, 52, 69, 88, 106–8, 248, 254
Prince George Agriplex, 96–102
Prince George Fire Centre, 81
Pugh, Trevor, 232

Quesnel, XIII, 1–2, 70–91, 103, 245, 254, 256
Quesnel Regional Airport, 83, 85

Radium, 4
Rankin, Rita, 165, 167
Ranta, John, 1, 24, 26–27
Raven Lake, 158
RCMP, 16, 20, 26–27, 30, 36–40, 46, **47**, 48–68, 85, 98, **106–7**, 108–11, **112**, 113–14, 134, 139–40, 151–52, 160–63, 171, 179–81, 197, 211–14, 220, 223, 230, 244, 251
Reid, Marcel, 49
Riske, L.W., 118
Riske Creek, III, VII, VIII, XVI, 1, 118–46, **147**, 148, **149**, 150–58, **159**, 160–65, 250
Rivett, Jim, 180–81
Roach, Jean, 6
Rock Creek, 245
Rocky Mountaineer, 21

INDEX

Roe Lake, 242
Rolf, Tim, 93, 96
Royce, 96
Rudy Johnson Bridge, 106

Safeway, 96, 114
Sage & Sands Trailer Park, 28
Salmon Arm, 155
Sandman evacuation centre, 107
Sanson, Jeff, 264
Save-On Foods, 114
Schreiner, Gord, 204–5
Schwartz, Don, 238
Scotty Creek, 41
Secwepemc First Nation, 216
Severin, Val, **48**, 56, 246
Sheridan Lake, **v**, 7, 8–14, 46, 51, 191, 200, 202, 226–51
Sheridan Lake Road West, 263
Sheridan West Forest Service Road, **xvi**, 263
Shields, Lana, 93–102, 248
Sieb, Jeremy, 1, 81–85, 90
Siegel, Ethan, 25
Sierra Club BC, 256
Simard, Suzanne, 261
Simpson, Roy, 38
Sims, Jennie, 203
Skeetchestin, 206
Skookumhorse Ranch, 174–75, **248**, 249
Skrepnek, Kevin, 188
Slater Mountain, **x**, **xi**, **xii**, **117**, 122–26
Smith, Cathy, 236
Smith, Court, 93
Smith, Lorne, 200–206

Smolen, Samantha ("Sam"), 116, **117**, 118, **119**, 120–27
Soda Creek, 105, 242
Soda Creek Road, 124
South Cariboo Search and Rescue (SAR), **48**, 49, 55–56, 246
South Green Lake, 49, 210–17
South Green Lake Volunteer Fire Department, 198, 217
Sperling, John, 209
Spokin Lake, 62, 104
Stack Valley Road, 149
Stewart, Nathan ("Stewy"), 84–86
Stone, Todd, 6
Sugar Cane, 67, 97, 105
Surrey, 111
Sweden, 253–55, 261, 264

Tatla Lake, 163
Taylor, Jeff, 162
Taylor, Mark, 73–74, 77, 79
Tegart, Jackie, 203
Terrace Fire Department, 142, 145
Tete Jaune, 3
Thain, Cat, 20
Thomas Lake, 241
Thompson, Steve, 4–5
Thompson River, 3, 169
Thompson-Nicola Regional District (TNRD), 170, 181, 187, 201, 210–11, 246
Three Corners, 10
Thuya Lake, 12
Timber Kings, 216
Tin Cup Lake, 197–198, 202, 214
Todd, Lee, **iii**, 122, **123**, 124–27, 158–63

Tolko Industries, 109
Toosey First Nation, 1, **2**, 120, 148–49, 250
Torch (dog), 241, **242**
Travers, Ray, 253–55
Tsilhqot'in language, 250

University of British Columbia, 261
University of Victoria, 253

Valley Drive, 171–72, 175
Van Ember, Brenda, 146, **147**, 150, 155
Van Ember, Chris, 147–50, **151**, 152–56
Van Ember, Kurt, 1, 146, **147**, 148–57
Vancouver, 114, 169, 177. *See also* North Vancouver
Vancouver Island, 189–90, 258
Vavenby, 254
Venables Valley, 17
Vernon, 199–202
Victoria, 4
Vieira, Hannah, 210–11, **212**, 213, 215
Vieira, Krista, 208–11, **212**, 213–15
Vieira, Makayla, 210, 215
Vieira, Miguel, 208–13
Vogt, Jeremy, 113–14, 244, 251
Volunteer Firefighters Association of BC, 272n27

Walch, Jason, 118, **119**
Wallach, Chelsea, 96–97, 102
Watch Lake, 230–31, 241
Watson, Sally, 210
West Fraser Mills, 205–6, 254
West Fraser Road complex, 81–82, 90, 245

West Fraser Timber, 109
Western Horse Review, 132
Whistler, 13
White, Josh, 16, **17**, 18–21
White Lake, 105–6, 109
White Rock, 68
Wholesale Club, 114
Wieting, Jens, 256–58
Wildwood, 49, 62, 66, 88, 94, 104–5
Williams, Bob, 254
Williams Lake, ii, x, 1, 7, 12, **25**, 48, 57, 61–63, 67, 69, 81–82, 87–88, 92–127, 130, 139–40, 147, 157, 162–63, 171, 190, 210, 213, 244, 248
Williams Lake Airport, iv, 62, 80, 104–5, 110–13, 245
Williams Lake City Hall, 109–13
Williams Lake Fire Department, 154
Williams Lake Stampede, 93–99, 105, 146
Wilson, Craig, 84–90
Wind and The Pillows, The, 216, **219**, 223
Windy.com, 237
Wohlleben, Peter, 260–61
Wood Pellet Association of Canada, 255
Woodburn, Barb, 36–43, 187, 189, 244
Woodburn, Rob, 36–43, 187, 244

Yale, 178, 242
Young Lake, 201–2, 205–6
YouTube, 235–36